Automatic Generation Of Algorithms

In the rapidly evolving domain of computational problem-solving, this book delves into the cutting-edge Automatic Generation of Algorithms (AGA) paradigm, a groundbreaking approach poised to redefine algorithm design for optimization problems. Spanning combinatorial optimization, machine learning, genetic programming, and beyond, it investigates AGA's transformative capabilities across diverse application areas. The book initiates by introducing fundamental combinatorial optimization concepts and NPhardness significance, laying the foundation for understanding AGA's necessity and potential. It then scrutinizes the pivotal Master Problem concept in AGA and the art of modeling for algorithm generation. The exploration progresses with integrating genetic programming and synergizing AGA with evolutionary computing. Subsequent chapters delve into the AGA-machine learning intersection, highlighting their shared optimization foundation while contrasting divergent objectives. The automatic generation of metaheuristics is examined, aiming to develop versatile algorithmic frameworks adaptable to various optimization problems. Furthermore, the book explores applying reinforcement learning techniques to automatic algorithm generation. Throughout, it invites readers to reimagine algorithmic design boundaries, offering insights into AGA's conceptual underpinnings, practical applications, and future directions, serving as an invitation for researchers, practitioners, and enthusiasts in computer science, operations research, artificial intelligence, and beyond to embark on a journey toward computational excellence where algorithms are born, evolved, and adapted to meet ever-changing real-world problem landscapes.

Victor Parada is Titular Professor in the Informatics Engineering Department at the University of Santiago, Chile.

Advances in Metaheuristics

Series Editors:
Patrick Siarry, *Universite Paris-Est Creteil, France*
Anand J. Kulkarni, *Symbiosis Center for Research and Innovation, Pune, India*

Metaheuristics for Enterprise Data Intelligence
Edited by Dr Kaustubh Sakhare, Dr Vibha Vyas, Dr Apoorva S Shastri

Handbook of AI-based Metaheuristics
Edited by Patrick Siarry and Anand J. Kulkarni

Metaheuristic Algorithms in Industry 4.0
Edited by Pritesh Shah, Ravi Sekhar, Anand J. Kulkarni, Patrick Siarry

Constraint Handling in Cohort Intelligence Algorithm
Ishaan R. Kale, Anand J. Kulkarni

Hybrid Genetic Optimization for IC Chip Thermal Control: with MATLAB® applications
Mathew V K, Tapano Kumar Hotta

Handbook of Moth-Flame Optimization Algorithm: Variants, Hybrids, Improvements, and Applications
Edited by Seyedali Mirjalili

Combinatorial Optimization Under Uncertainty: Real-life Scenarios in Allocation Problems
Edited by Ritu Arora, Prof. Shalini Arora, Anand J. Kulkarni, Patrick Siarry

AI-based Metaheuristics for Information Security in Digital Media
Edited by Apoorva S Shastri, Mangal Singh, Anand J. Kulkarni, Patrick Siarry

Graph Coloring
Maurice Clerc

Automatic Generation Of Algorithms
Victor Parada

For more information about this series please visit: https://www.routledge.com/Advances-in-Metaheuristics/book-series/AIM

Automatic Generation Of Algorithms

Victor Parada

CRC Press
Taylor & Francis Group
Boca Raton London New York

CRC Press is an imprint of the
Taylor & Francis Group, an **informa** business

Designed cover image: shutterstock

First edition published 2025
by CRC Press
2385 NW Executive Center Drive, Suite 320, Boca Raton FL 33431

and by CRC Press
4 Park Square, Milton Park, Abingdon, Oxon, OX14 4RN

CRC Press is an imprint of Taylor & Francis Group, LLC

ISBN: 978-1-032-89445-4 (hbk)
ISBN: 978-1-032-89138-5 (pbk)
ISBN: 978-1-003-54289-6 (ebk)

DOI: 10.1201/9781003542896

Typeset in Minion
by SPi Technologies India Pvt Ltd (Straive)

To my wife Lorena and family and
my sons Leandro and Lucas.

Contents

Preface

THIS BOOK AIMS TO introduce the Automatic Generation of Algorithms (AGA), a concept emerging from integrating different ways to tackle challenging optimization problems and to provide a comprehensive framework that bridges the gap between theoretical research and practical applications. The classic approach to algorithm design involves creating a single, effective algorithm capable of solving any instance of a given problem. However, this is often impractical due to the highly complex and challenging landscape of all possible feasible solutions for specific problems. Metaheuristics, while powerful, introduce additional challenges, such as the necessity of meticulous parameter tuning to achieve optimal performance. The demand for algorithms that can adapt and solve any problem instance further complicates the process. These combined challenges highlight the significant gap in traditional methods and underscore the need for the AGA that can simultaneously search through all involved search spaces, including the space of instances, parameter configurations, and algorithmic strategies.

Traditional algorithm design approaches, while robust, can be significantly enhanced to better address the nuanced and dynamic nature of contemporary optimization problems through innovative solutions like the AGA. AGA emerges as a beacon of innovation, offering a framework for generating algorithms that are not only tailored to specific problem instances but also capable of evolving in response to changing conditions and requirements. This paradigm shift integrates advanced techniques from various domains, including machine learning, genetic programming, evolutionary computing, and metaheuristics, to systematically create and refine algorithms. AGA's framework supports continuous learning and adaptation, enabling the automatic refinement of algorithms based on real-time feedback and performance metrics. By addressing the limitations of traditional methods and incorporating a flexible, adaptive approach to algorithm design, AGA represents an advancement in computational optimization.

AGA is deeply rooted in the foundational principles of genetic programming. However, at its core lies a master optimization problem that extends far beyond this initial scope. By framing AGA as an overarching optimization challenge, we uncover many possibilities for creating new algorithms. This master problem involves minimizing the error between a newly generated algorithm and the optimal solution for a combinatorial optimization problem. This emerging field results from many existing scientific contributions from different views and various authors. As we solve this optimization problem, we produce a new algorithm and advance the frontier of algorithmic design, unlocking innovative approaches to tackle complex computational challenges.

This book aims to advance computational problem-solving, opening the path for a future where algorithms are not just static creations but dynamic entities capable of evolving to meet the demands of an ever-changing world. At its core, the book invites readers to reimagine the boundaries of algorithmic design. It addresses researchers, practitioners, and enthusiasts in computer science, operations research, artificial intelligence, and beyond. Whether you seek to deepen your understanding of AGA, explore its applications, or contribute to its evolution, the insights and discussions presented aim to inspire and inform. By bringing together diverse perspectives, the book fosters a deeper understanding of AGA's capabilities and inspires further research and development in this promising field.

As we stand on the brink of a new era in computational problem-solving, AGA offers a path forward that is both revolutionary and inevitable. It challenges us to reconsider our approach to algorithm design, advocating for a future where algorithms are not merely created but born, evolved, and adapted to meet the ever-changing landscape of our world's problems. Welcome to the journey of Automatic Generation of Algorithms—a journey toward the future of computational excellence

Acknowledgments

I WANT TO EXTEND MY deepest gratitude to my students, whose dedication and intellectual curiosity have constantly inspired me. Supervising Master's theses, Engineering Memories, and the PhD thesis has been a rewarding experience that has significantly contributed to the genesis of this book. Throughout our many discussions, the insights and challenging questions have been instrumental in shaping the ideas presented here. The collaborative efforts to improve the performance of AGA, using the machines available at various stages of technological advancement, have been particularly fruitful. The commitment and enthusiasm have greatly enriched this journey, and I am honored to have been part of your academic development.

The University of Santiago, Chile's sabbatical program enabled this work by allowing external research collaborations. I express sincere gratitude for this support. During my sabbatical, I conducted part of my research at the Pontifical Catholic University of Rio de Janeiro, thanks to an invitation from the Industrial Engineering Department promoted by my friend Emeritus Professor José Eugenio Leal. The stimulating environment of Rio de Janeiro contributed significantly to the development of several concepts presented in this book. Subsequently, I continued my research at Trinity University, San Antonio, TX, USA, with the generous support of my friend Robert Scherer, Dean of the Michael Neidorff Business School. My time in San Antonio was both productive and enjoyable. Furthermore, I acknowledge the invaluable contributions of Chile's Complex Engineering Systems Institute to this research. Their support across multiple research endeavors has been instrumental in advancing the work presented herein.

Structure of the Book

The book's organization ensures that readers progressively build their knowledge from foundational concepts to advanced applications, ultimately gaining a deep and practical understanding of AGA's potential and implementation. The journey begins with a foundational overview of optimization, setting the stage with key concepts such as combinatorial optimization and NP-hardness (Chapter 1). These initial discussions are crucial as they lay the groundwork for appreciating the complexities involved in optimization problems. By introducing the significance of NP-hardness, the book helps readers understand why specific problems are particularly challenging and why traditional algorithms often fall short. This chapter establishes a general framework for combinatorial optimization problems and explores various methods for tackling these challenges. It concludes by emphasizing the potential of AGA for algorithm generation, making it a pivotal tool for both practical applications and enhancing educational understanding of optimization concepts.

Building on this foundation, the book delves into the Master Problem, a central theme in AGA (Chapter 2). This chapter addresses the complexities of searching the spaces of algorithms, parameters, and problem instances to maximize algorithm performance while minimizing errors. It emphasizes the need for tailored approaches in algorithm design and explores the No-Free-Lunch theorem and its implications. The chapter simultaneously and independently covers parameter tuning and the algorithm selection problem, providing a deep understanding of the interplay between algorithms, their parameters, and the problem space. This thorough exploration sets the stage for appreciating the innovative potential of AGA in navigating these multifaceted challenges.

Chapter 3 focuses on modeling, particularly for NP-hard optimization problems in Operations Research (OR) and Artificial Intelligence

(AI). It outlines a structured methodology for identifying and formulating real-world problems into mathematical models. This chapter highlights the versatility of AGA in adapting to diverse, complex scenarios by bridging theoretical approaches with practical applications. It explores fundamental models in OR and AI, demonstrating how AGA can significantly enhance problem-solving capabilities through customized algorithms.

The book then introduces the concept of AGA within the genetic programming (GP) framework (Chapter 4). This section begins with a historical perspective on the quest for self-programming machines, tracing the evolution of thought leading to John Koza's formalization of GP. It delves into evolutionary computing, genetic operations, and the modeling of the Master Problem using GP. By showcasing how GP can automatically generate specialized algorithms tailored to specific classes of problem instances, the book illustrates a transformative approach to algorithmic design. This chapter highlights the synergy between AGA and evolutionary computing, demonstrating the power of combining these methodologies to create highly effective optimization algorithms.

The book explores the relationship between AGA and machine learning, focusing on optimization as a pivotal concept bridging both domains (Chapter 5). This chapter discusses fine-tuning mathematical models to datasets and the principle of meta-optimization in AGA. It contrasts overfitting in machine learning with specialization in AGA, highlighting their different objectives. The exploration emphasizes how AGA can generate highly efficient, problem-specific algorithms by comprehensively exploring algorithmic components and configurations. This nuanced discussion underscores the multifaceted roles of optimization in computational problem-solving, bridging theoretical concepts with practical applications in algorithm generation.

The book examines the automatic generation of metaheuristics, aiming to design algorithms with metaheuristic characteristics without extensive human expertise (Chapter 6). It explores the key concepts in metaheuristics, such as solution representation, neighborhood exploration, and the roles of solution containers and terminals. The chapter underscores the potential of automatically generating metaheuristics to solve complex optimization problems, highlighting this approach's adaptability and broad applicability. By autonomously navigating the vast array of metaheuristic components, AGA identifies and synthesizes those that promise enhanced performance across diverse optimization challenges.

The final chapter examines the application of reinforcement learning (RL) to AGA. It begins with the origins of RL in psychology, optimal control, and dynamic programming, providing a brief historical context for its development. The chapter discusses Bellman's Principle of Optimality and its role in developing dynamic programming algorithms. It frames the AGA challenge within the Deciding Agent Problem context, illustrating how RL algorithms can be effectively modeled and applied to this domain. The book reveals a promising avenue for advancing computational problem-solving methodologies by exploring the potential of combining AGA with RL.

CHAPTER 1

Overview of Optimization

1.1 INTRODUCTION

Automatic generation of algorithm (AGA) is a technique used to construct new algorithms through automatic exploration and a combination of existing algorithmic components. The AGA constructs the best algorithm for an optimization problem with greater specificity from primitive elements that typically compose an algorithm. The construction process occurs by solving an optimization meta-problem to minimize the algorithm error relative to the correct value of a set of problem instances. The optimization meta-problem, which we call the master problem, aims to minimize the error by searching the space containing the possible algorithms for the problem. Consider an initial randomly generated algorithm for an optimization problem capable of finding a solution to such an optimization problem. Consider further that a set of instances of the optimization problem with a known optimal solution value is available. Then, the initial algorithm can eventually produce a non-optimal solution for a specific problem instance. Consequently, an objective function value difference arises between the optimal solution and the solution found for this algorithm, called the error.

The goal of AGA is to find the algorithm that provides a solution that is close to the optimum and simultaneously is human-readable and does not exceed a predefined size. Although any optimization technique can approach

DOI: 10.1201/9781003542896-1

the optimization master problem, existing studies have preferentially explored genetic programming to approach this problem. Genetic programming is an evolutionary algorithm that searches the solution space by emulating the Darwinian laws of natural evolution. GP uses syntax trees to represent a solution, which are ideal data structures for an assembly of instructions such as those occurring in an algorithm. For this reason, genetic programming is the preferred technique for automatically generating algorithms.

In AGA, the process begins with formulating the master problem. This means identifying the objective function and the feasible space for searching for a solution that, in turn, is an algorithm. Thus, a set of algorithmic components that can be combined to create algorithms should be defined. These components can be parts of existing heuristics, complete heuristics, or even exact methods and, when combined, result in a syntax tree. Thus, an initial population of syntax trees is necessary when an evolutionary algorithm is used to solve the meta-optimization problem. The evolutionary process consists of applying genetic operators such as reproduction, crossover, and mutation to the population of algorithmic trees. These operators mimic the processes of natural selection and genetic variation. The population evolves over several generations, and the fitness of each algorithmic tree is evaluated based on its performance in solving the problem. During the evolution process, the method explores the space of feasible algorithms by generating new algorithmic combinations and selecting those that show promising performance. The objective function evaluates each new algorithm using a set of problem instances with known optimal solutions. The master problem objective function may use various performance measures, such as mean square or absolute error.

AGA's core lies in framing the creation of algorithms to solve the master problem. This approach is powerful because it shifts the focus from manually designing algorithms to defining the space and criteria for algorithmic evolution. This meta-problem encompasses specifying an objective function and a feasible search space, essentially defining what constitutes a "good" algorithm for a given problem and how to explore the space of potential solutions. The AGA method opens a vast field of possibilities. Automating the process of algorithm generation can potentially discover new, efficient algorithms that might not be intuitive to human designers. However, this approach also comes with challenges, such as ensuring sufficient diversity in the population of algorithms, avoiding local optima, and managing the computational complexity of evaluating and evolving algorithmic trees. While genetic programming is a natural fit to solve the meta-optimization problem, the AGA method is not limited to it. The same master problem can be tackled using other metaheuristic methods like particle swarm optimization and ant colony optimization or even approaches like reinforcement learning in machine learning. Each method brings its strengths and biases to the algorithm generation process, potentially leading to different solutions. AGA represents a cutting-edge intersection of algorithmic design, evolutionary computation, metaheuristic optimization, and artificial intelligence, with promising applications in various problem domains. AGA is a perfect fit for combinatorial optimization problems since the search for novel algorithms arises from the necessity to overcome the NP-hardness.

1.2 COMBINATORIAL OPTIMIZATION

Combinatorial optimization is a field of mathematics and, more specifically, optimization. It studies how to approach optimization problems in which an algorithm must find the best solution among a finite set of combinations of discrete objects. These problems involve choosing among a finite set of possibilities to find the best or optimal solution according to a specific objective or criterion. In combinatorial optimization, the set of feasible solutions is usually a combinatorial structure such as graphs, networks, permutations, subsets, or sequences. The main characteristic of combinatorial optimization problems is that the search space grows exponentially with the problem's size, making them computationally challenging. Finding the optimal solution from the many possible combinations requires efficient algorithms and heuristics to explore the search space. Thus, combinatorial optimization ranges from the mathematical modeling of a real-world situation to selecting or designing the appropriate algorithm to find the optimal or near-optimal solution.

Combinatorial optimization problems arise in decision-making in various fields and applications, such as operations research, logistics, scheduling, telecommunications, finance, computer science, artificial intelligence, and engineering. There are countless applications of combinatorial optimization in management. Some of the best-known combinatorial optimization problems are the traveling salesman problem (TSP), the knapsack problem, the graph coloring problem, the maximum flow problem, the minimum tree problem, and the assignment problem. These theoretical problems are appropriate for representing many real-world situations. For example, modelers often use the TSP as a subproblem for modeling vehicle routing in urban, maritime, and air transportation.

Combinatorial optimization is crucial in decision-making and resource allocation in various industries and domains. By finding optimal or near-optimal solutions to combinatorial optimization problems, organizations can optimize their processes, allocate resources efficiently, minimize costs, improve scheduling, and increase overall performance. Consequently, combinatorial optimization is a field in most organizations because it is always necessary to use scarce resources best.

Effective solving of combinatorial optimization problems usually involves the design of algorithms and heuristics that exploit the structure and properties of the problem. Exact methods, such as integer or dynamic programming, can be employed to find optimal solutions for small problem instances. However, large-size problems require many computational resources, making constructing a user-oriented computational system unfeasible. Approximate algorithms, metaheuristics, and hybrid techniques often allow us to find near-optimal solutions in a reasonable time. Difficulty Problems form the basis of scientific research in combinatorial optimization. The field of AGA arises in this area. Since there are many possibilities for combining existing algorithms to determine the best way to solve a given optimization problem, it seems appropriate to search this space of possibilities automatically.

1.3 NP-HARDNESS

Computational complexity theory is a branch of theoretical computer science that deals with the resources required during computation to solve a given problem. The most common resources are time, typically measured as the number of steps to solve a problem, and space, measured as the computer memory required. Thus, a problem is said to be solvable in polynomial time if the time it takes to solve the problem can be expressed as a polynomial function of the input size for the problem. For instance, an algorithm that takes time proportional to n^2, where n is the input size, is polynomial. Polynomial time is considered "efficient" or "feasible" in computation. In addition, we define the NP (Nondeterministic Polynomial-Time) class as one where, given a proposed solution, one can verify its correctness in polynomial time. The term "nondeterministic" refers to the concept of guessing a solution and then verifying its correctness. Conversely, a problem is NP-hard if one can transform every problem in NP into this problem in polynomial time. Essentially, NP-hard problems are at least as challenging as the hardest problems in NP. However, unlike NP problems, an NP-hard problem might not have a verifiable solution in polynomial time.

The process of transforming one problem into another in such a way that the algorithm for the second problem can be used to solve the first problem, and this transformation process takes polynomial time, is known as polynomial time reduction (PTR). Thus, PTR is a method of converting one problem (let's call it π_A) into another problem (π_B) so that a solution to π_B, can be used to solve π_A, and the conversion process takes polynomial time. This method is crucial for comparing the relative difficulties of different computational problems. It is a fundamental concept in computational complexity theory, particularly in studying the relationship between P, NP, and NP-hard complexity classes.

Having defined classes of problems arises the most important open problem in computer science: if P = NP? It asks whether every problem whose solution can be verified quickly (in polynomial time, i.e., in NP) can also be solved quickly (again in polynomial time, i.e., in P). If P were equal to NP, it would imply that every problem whose solution can be quickly verified could also be quickly solved.

NP-hard problems are often computationally complex to solve exactly, which motivates the development of approximation algorithms and heuristics to find near-optimal solutions within reasonable time bounds. The concept of NP-hardness indicates that exactly solving a problem is computationally challenging. It suggests that there may not be a polynomial algorithm to solve the problem in the general case. In practice, the only viable approach for large-size cases is to use approximation algorithms, heuristics, or other techniques to find near-optimal solutions. Although NP-hardness indicates the difficulty of solving a problem exactly, it does not mean that all instances of an NP-hard problem are equally tricky. Some instances may have specific features that enable efficient solutions, and researchers and practitioners often focus on developing specialized algorithms and heuristics to exploit these features.

1.4 NP-HARDNESS IN COMBINATORIAL OPTIMIZATION

Combinatorial optimization and NP-hardness are closely related concepts in the field of computational complexity theory. Specifically, NP-hardness is an indication of the computational complexity of a problem. Many significant combinatorial optimization problems are known to be NP-hard. Some of the best-known NP-hard combinatorial optimization problems are the TSP, the knapsack problem, the graph coloring problem, the maximum independent set problem, and many others. These problems have critical real-world applications but are known for their computational difficulty. Problems arise in various domains and industries and often involve finding the best configuration, allocation, arrangement, or assignment of resources or elements. Because of NP-hardness, solving large instances of NP-hard combinatorial optimization problems requires using approximation algorithms, heuristics, or other techniques to find near-optimal solutions within reasonable time limits. These methods do not guarantee the determination of the optimal solution but only allow for finding an approximate solution.

The practical implications of NP-hard combinatorial optimization problems are significant and wide-ranging across various industries and domains. For example, combinatorial optimization problems are the core of making decisions in logistics, transportation, and manufacturing industries. Solutions to problems like the TSP or the knapsack problem can lead to more efficient routing, better use of storage space, and optimal inventory management, thereby reducing costs and improving operational efficiency. The computational resources required to find an optimal solution can be prohibitive for large-scale instances. Thus, it is necessary to use powerful computers and sophisticated algorithms, leading to significant investments in computational infrastructure. These methods quickly provide "good enough" solutions, enabling businesses to make timely decisions. However, the trade-off is that these solutions may not be optimal, which could lead to less efficient outcomes compared to the theoretical best. Consequently, the quality of solutions obtained from these algorithms directly impacts strategic and operational business decisions. For example, in supply chain management, decisions based on these algorithms affect delivery routes, inventory levels, and production schedules.

1.5 GENERAL FRAMEWORK OF A COMBINATORIAL OPTIMIZATION PROBLEM

A general mathematical statement of a combinatorial optimization problem is: Let $f(x)$ be an objective function that needs to be minimized or maximized, and x is a vector of decision variables representing the choices, allocations, or configurations of the problem. The objective function $f(x)$ quantifies the measure of the quality or desirability of a solution. Also, consider a set of constraints that bound the feasible solutions denoted as Ω. The goal is to find the values of the decision variables x that minimize or maximize the objective function while satisfying all the constraint space Ω. The specific form of the objective function and constraints will depend on the problem at hand. The decision variables x can take on binary values (0 or 1), belong to a finite set of values, or be continuous variables. The objective function and constraints can involve linear functions, nonlinear functions, or a combination.

$$\min f(x) \text{subject to} : x\Omega \tag{1}$$

The general mathematical statement provides a framework to express various combinatorial optimization problems. The challenge lies in formulating the specific objective function and constraints that capture the problem's characteristics and requirements. This framework is also helpful in representing many practical and theoretical situations and facing such problems has given rise to many fields. It is important to note that the borders of these fields are not completely clear, and some problems can be considered to belong to two or more fields. Examples include:

- Linear Programming. The objective function and each constraint are linear.
- Nonlinear Programming: The objective function is nonlinear, and the constraint space can be linear or nonlinear.
- Integer Programming: The objective function is typically linear, though it can sometimes be nonlinear. The constraint space is linear, with the additional constraint that some or all variables are integers.
- Mixed-Integer Programming: The objective function is linear or nonlinear, and the constraint space is linear or nonlinear, with some variables constrained to be integers.
- Quadratic Programming: The objective function is quadratic, and the constraint space is linear.
- Convex Optimization: The objective function is convex, a particular case of nonlinear and convex constraint space.
- Combinatorial Optimization: The objective function is often linear but can be nonlinear. The constraint space is discrete and involves a finite or countably infinite set.
- Stochastic Programming: The objective function can be linear or nonlinear, and the constraint space involves probabilistic or stochastic elements.

- Dynamic Programming: The objective function can vary, and the constraint space is composed of problems broken down into simpler subproblems, often involving time or stages.
- Global Optimization: The objective function is nonlinear, and the constraint space can be linear or nonlinear.
- Machine Learning: This fundamental field of artificial intelligence focuses on developing algorithms to learn from data and make predictions or decisions. The optimization process typically involves the minimization of a loss function to improve model accuracy.

Those different fields are one of humanity's most potent tools for addressing real-world challenges. Its broad applicability spans numerous industries, facilitating efficient resource allocation, cost reduction, and enhanced decision-making. In logistics, for instance, it streamlines complex routing and scheduling tasks, while in finance, it aids in portfolio optimization. The innovation spurred by the quest to solve intricate optimization problems has significantly advanced computational methods and algorithms, contributing to technological and scientific progress. Moreover, the societal impacts of optimization are profound, evident in improved healthcare systems, sustainable environmental management, and optimized public services. This interdisciplinary field, bridging mathematics, computer science, engineering, and beyond, not only addresses current challenges but also equips us with the tools and knowledge to tackle future complexities. Optimization techniques' continuous evolution and refinement ensure their enduring relevance and capacity to contribute meaningfully to various facets of modern life.

1.6 METHODS FOR COMBINATORIAL OPTIMIZATION PROBLEMS

Various methods can search for the optimal solution of combinatorial optimization problems. Such methods are in some of the following classes of methods.

- Exact methods: These are methods that determine the optimal solution to a combinatorial optimization problem by potentially exploring the entire solution space. These methods ensure the determination of the best solution but may be computationally infeasible for large-size problems. Examples of exact methods are branch and bound, branch and Price, and cutting plane methods.

- Heuristic methods: Heuristic methods are approximate algorithms that prioritize efficiency over optimality. Such methods aim to find near-optimal solutions in a reasonable time. Depending on the optimization problem, in some cases, they can determine the optimal solution; however, no theoretical certificate guarantees the determination of the optimal solution. A heuristic uses problem-specific techniques and strategies to guide the search toward promising areas of the solution space. For this reason, using heuristics in conjunction with an exact method is possible, i.e., the heuristic determines an initial solution, and then the exact method is applied. Examples of heuristic methods are constructive heuristics that iteratively construct a solution to the problem, local search, and nearest neighbor heuristics.
- Metaheuristic methods: These are high-level search strategies that guide the exploration of the solution space. So, they are general-purpose optimization techniques practical for various combinatorial optimization problems. Metaheuristic methods consider iterative neighborhood exploration processes to seek an improvement in the current solution. Some examples of metaheuristic methods are tabu search, simulated annealing, genetic algorithms, ant colony optimization, particle swarm optimization, iterated local search, and variable neighborhood search. Many new metaheuristics have appeared during the last two decades, mimicking diverse natural phenomena. These include algorithms inspired by the behavior of fireflies, cuckoo search, and the gravitational interactions among celestial bodies, reflecting the continuous innovation in the field.
- Approximation algorithms: An approximation algorithm determines a solution guaranteed to be within a specific deviation range from the optimal solution. These algorithms offer a trade-off between solution quality and computational time. Approximation algorithms usually exploit the structures and properties of the problem to find efficient and reasonably good solutions. Examples of approximation algorithms are greedy algorithms and linear programming relaxations.
- Hybrid methods: Hybrid methods combine different approaches and techniques to exploit their strengths and mitigate their weaknesses. These methods aim to achieve better quality and efficiency of solutions by integrating multiple algorithms or problem-specific strategies. Hybrid methods often combine elements of exact methods, heuristics, metaheuristics, and approximation algorithms.

- Problem-specific methods: Some combinatorial optimization problems have unique characteristics that allow the development of specialized algorithms for their structures. These problem-specific methods use domain knowledge, problem-specific heuristics, or mathematical properties to find efficient solutions. Examples include specialized algorithms for the TSP, the vehicle routing problem, or the graph coloring problem.

The AGA represents a groundbreaking advancement in combinatorial optimization, offering the potential to amalgamate all existing optimization techniques and, more significantly, to innovate new algorithms previously undiscovered by experts. By harnessing this approach, we can integrate the precision of exact methods, the efficiency of heuristic and metaheuristic strategies, the reliability of approximation algorithms, and the tailored effectiveness of problem-specific methods. Additionally, the potential for hybridizing these diverse approaches opens up possibilities for more efficient solutions that are adaptable to the complex and varying nature of real-world problems. This automated, holistic framework for algorithm generation offers a new perspective in the optimization landscape, pushing the boundaries of what can be achieved in computational problem-solving and offering a new horizon of possibilities in tackling some of the most challenging problems across various domains.

1.7 A GENERAL OPTIMIZATION ALGORITHM

An algorithm is a step-by-step procedure that considers a set of instructions to perform a specific task. An algorithm produces an output from input data to achieve a given goal. Algorithms are fundamental in computer science and are crucial in problem-solving, data processing, information retrieval, and computational tasks. They serve as building blocks for software development, providing the logic and instructions for computers to perform specific tasks efficiently and accurately.

Algorithms are procedures that describe the actions and calculations required to solve a problem. They provide a systematic and structured problem-solving approach by breaking down complex tasks into smaller, manageable, and programmable steps in a computer language. Thus, an algorithm must have precise and unambiguous instructions that can be followed unambiguously. Each step of the algorithm must be well-defined and clearly understandable. Likewise, it must be designed to terminate in a finite number of steps, avoiding infinite loops or infinite execution. In addition, each step of an algorithm must be deterministic, meaning that given the same input, the algorithm must always produce the same output. It must not involve randomness or unpredictable behavior. Algorithms can be expressed using various forms of representation, such as natural language, pseudocode, syntax trees, flowcharts, or programming languages. An algorithm must have well-defined inputs and outputs. From the initial data or values on which the algorithm operates, which may be external or obtained in previous algorithm steps, the algorithm produces an output or outputs, which are the results or solutions generated by the algorithm after performing the specified calculations or operations.

An optimization algorithm is a set of well-defined instructions for finding a solution to a specific optimization problem. Thus, optimization methods are specified in detail by an optimization algorithm. In the context of optimization, "optimization algorithm" and "optimization method" are terms often used interchangeably to refer to a computational technique or approach used to solve optimization problems. However, we identify an optimization algorithm as a specific computational procedure or technique used to find the optimal solution within an optimization problem. On the other hand, an optimization method is considered a broader term encompassing various techniques, approaches, and frameworks used to solve optimization problems, often involving a combination of algorithms and problem-specific techniques. Optimization algorithms have several characteristics defining their behavior and effectiveness in solving optimization problems. It is appropriate for an optimization algorithm to contain the following components:

- Objective function: Optimization algorithms operate with an objective function that quantifies the quality measure of a solution. The algorithm searches for the optimal solution by optimizing this objective function based on the problem's constraints.
- Solution space: Optimization algorithms explore the solution space, which contains all possible solutions to the optimization problem. The size and structure of the solution space depend on the problem variables, the constraints, and the specific characteristics of the problem.
- Initialization stage: Optimization algorithms usually require an initial solution belonging to the solution space. The choice of the initial solution can influence the algorithm's convergence and the final solution's quality. Different initialization strategies, such as random initialization, problem-specific techniques, or other heuristic algorithms, can be used.

- Search strategy: An optimization algorithm employs a search strategy to navigate the solution space to find better solutions iteratively. The search strategy determines how the algorithm explores the solution space, moves from one solution to another, and converges to an optimal or near-optimal solution. Common search strategies include local, global, gradient-based, evolutionary, or neighborhood exploration.
- Iterative improvement cycle: Optimization algorithms usually work iteratively, repeatedly modifying and refining the current solution to improve its quality. Therefore, the algorithm maintains memories in which it stores the current solution and other solutions already produced that, in some way, facilitate the search. They iteratively update the solution based on predefined rules or algorithms until it reaches convergence or a termination condition. Each iteration approaches the optimal solution, considering the objective function and the problem's constraints.
- Convergence criteria: Optimization algorithms require convergence criteria to determine when to stop the iterative process. Convergence criteria may be based on reaching a certain solution quality, a specific number of iterations, or a tolerance level indicating that further improvement is unlikely to affect the solution significantly.
- Constraint management: To ideally perform the search within the solution space, optimization algorithms manage constraints by considering them during the search process. It is also possible for an algorithm to perform the search by visiting some infeasible solutions and resorting to penalty functions that penalize the value of the objective function of infeasible solutions.
- Solution representation: The modeling of an optimization problem must consider a representation scheme of the feasible space. The choice of representation depends on the characteristics of the problem and can be binary, real-valued, categorical, or a combination of these. Then, the corresponding optimization algorithm must consider a representation scheme to encode the solutions within the solution space. The representation determines the structure and operations applicable to the solutions during the search process.
- Performance metrics: The effectiveness of optimization algorithms is evaluated using performance metrics. A modeler may consider various metrics such as solution quality, convergence speed, computation time,

scalability to large problem instances, robustness to noise or uncertainty, and ability to handle complex or multimodal solution spaces.

Algorithm 1.1 is a general framework representing most combinatorial optimization algorithms. This pseudocode provides a flexible template, which can be tailored to various specific problems and requirements. This general optimization algorithm framework plays a pivotal role in combinatorial optimization, serving as a versatile blueprint encompassing a wide array of optimization fields. Its significance lies in its ability to provide a cohesive overview of diverse optimization methodologies, making it an invaluable tool for practical problem-solving across various industries. By integrating key components such as objective functions, solution space exploration, and iterative improvement strategies, this framework adapts to many complex real-world scenarios, ranging from logistics and supply chain management to network design and resource allocation. This holistic approach bridges theoretical knowledge with practical application and fosters innovation and creativity in tackling the most demanding combinatorial optimization problems in academic and professional settings.

Algorithm 1.1: General Combinatorial Optimization Algorithm

```
Input: Problem _ instance;
Output: Optimal_solution;
Begin
   Define ObjectiveFunction(solution);
   Define SolutionSpace(Problem_instance)
   current_solution = InitializeSolution(SolutionSpace);
   best_solution = current_solution;
   while not Convergence_Criteria_Met(best_solution,
     iteration_count, tolerance)
      new_solution = SearchStrategy(current_solution,
        SolutionSpace);
      if IsFeasible(new_solution, problem_instance.
        constraints)
        new_solution = HandleConstraints(new_solution);
        if ObjectiveFunction(new_solution) better than
          ObjectiveFunction(best_solution)
            best_solution = new_solution;
            current_solution = new_solution;
        iteration_count += 1;
   optimal_solution = best_solution;
   return optimal_solution
End.
```

This optimization algorithm methodically addresses complex problem-solving through a series of integral components. It begins by defining an "ObjectiveFunction" tailored to quantify solution quality, guiding the optimization process. The "SolutionSpace" based on the problem instance, encompasses all potential solutions, setting the boundaries for exploration. The algorithm initiates with a "current_solution" typically derived from heuristic methods, establishing a starting point for the search; simultaneously, this solution is also considered the initial "best_solution" for comparison. Central to the algorithm is the iterative improvement cycle, where a "SearchStrategy" dynamically navigates through the solution space, continually seeking improvements. Each iteration involves checking the feasibility of new solutions against predefined constraints ("IsFeasible") and adjusting them as necessary ("HandleConstraints"). The algorithm compares new solutions to the current best, updating it only if a better solution is found. A convergence criterion governs this process, typically based on solution quality, iteration count, or improvement threshold, ensuring the algorithm terminates once an optimal solution is unlikely to be further refined. Ultimately, the "best_solution" identified through this rigorous and iterative process is returned as the "optimal_solution", exemplifying a well-rounded approach that balances exploration and exploitation in the quest for optimality.

1.8 SUMMARY

This chapter outlines the AGA, a technique for constructing new algorithms by combining existing components to solve optimization problems efficiently. It introduces combinatorial optimization, detailing its significance in solving real-world problems by finding optimal solutions among discrete objects. The chapter delves into NP-hardness, explaining its relevance in computational complexity and its challenge in finding solutions. It presents a general framework for combinatorial optimization problems, highlighting various methods for tackling these, including exact, heuristic, metaheuristic, approximation algorithms, hybrid methods, and problem-specific approaches. The chapter concludes by emphasizing AGA's potential in revolutionizing algorithm generation, making it a pivotal tool for both practical applications across industries and enhancing educational understanding of optimization concepts.

EXERCISES

1. Reflect on how optimization techniques are applied across various logistics, finance, healthcare, and telecommunications fields. Provide specific examples of optimization problems in each field, such as vehicle routing in logistics, portfolio optimization in finance, patient scheduling in healthcare, and network design in telecommunications. Discuss how these examples illustrate the transversal nature of optimization.

2. Explain the concept of NP-hardness and why it poses a significant challenge in solving large-scale combinatorial optimization problems. Provide examples of NP-hard problems. Discuss the implications of NP-hardness for practical problem-solving and how approximation algorithms and heuristics are used to address these challenges.

3. Identify and describe optimization problems that occur in the operative management of a health center, including both machine learning and operations management optimization. Consider patient scheduling, staff rostering, resource allocation, inventory management, appointment booking, treatment planning, and emergency response optimization. Additionally, explore machine learning optimization problems such as predictive modeling for patient no-shows, disease outbreak prediction, personalized treatment recommendations, and demand forecasting for medical supplies. Explain how each identified problem can be modeled as an optimization problem, detailing the decision variables, objective functions, and constraints involved. Analyze the NP-hardness of each problem.

4. Identify and describe optimization problems that occur in city transportation management, including machine learning optimization and operations management optimization. Consider traffic flow optimization, public transportation scheduling, route planning for buses and trains, ride-sharing and taxi dispatching, parking space allocation, and infrastructure maintenance scheduling. Additionally, explore machine learning optimization problems such as predicting traffic congestion, optimizing dynamic pricing for ride-sharing services, demand forecasting for public transportation, and real-time route optimization based on traffic conditions. Explain how each identified problem can be modeled as an optimization problem, detailing the decision variables, objective functions, and constraints involved.

5. Identify and describe optimization problems in managing a retail company, including machine learning and operations management optimization. Consider aspects such as inventory management, demand forecasting, pricing optimization, staff scheduling, supply chain optimization, layout optimization of retail spaces, and delivery routing. Additionally, explore machine learning optimization problems such as customer segmentation for targeted marketing, recommendation systems for personalized product suggestions, predictive analytics for sales trends, and dynamic pricing based on real-time demand. Explain how each identified problem can be modeled as an optimization problem, detailing the decision variables, objective functions, and constraints involved.

6. Describe the AGA framework and its significance in solving complex optimization problems. Provide an example of how AGA can be applied to a specific problem, such as scheduling tasks in a manufacturing plant or optimizing delivery routes. Discuss the benefits and challenges of using AGA compared to manually designing algorithms.

7. Choose a well-known combinatorial optimization problem (e.g., Graph Coloring Problem) and describe a heuristic method to solve it. Define the problem, the heuristic approach, and its steps. Discuss the advantages and limitations of using heuristic methods for solving combinatorial optimization problems, particularly regarding solution quality and computational efficiency.

8. Discuss hybrid optimization methods and their advantages in solving complex problems. Provide an example of a hybrid approach combining exact methods and metaheuristics. Explain how this hybrid approach can improve solution quality and computational efficiency compared to a single method.

CHAPTER 2

The Master Problem

2.1 INTRODUCTION

The realm of optimization offers various approaches for representing real-world practical problems, often captured through mathematical optimization formulations. These mathematical models are powerful tools to mimic real-world scenarios, aiding in designing and implementing practical solutions. However, when dealing with specific optimization problems, selecting the most efficient algorithm to uncover practical solutions becomes complex. This complexity arises when the problem belongs to a challenging class where no readily available algorithm can efficiently solve it. Consequently, even when a real-world situation aligns perfectly with a recognized optimization problem, the quest for a practical solution becomes intertwined with the search for the optimal algorithm to tackle it.

We can use the optimization field to identify the best-suited algorithms in such intricate scenarios. This use implies that the pursuit of the ideal algorithm can also be modeled as an optimization problem in its own right. This problem, the master problem, involves a comprehensive exploration of the space encompassing all conceivable algorithms to select the algorithm that ultimately yields the most exceptional performance. By treating the algorithm selection as an optimization challenge, it becomes possible to navigate the intricate landscape of complicated problems and ascertain the algorithm that maximizes the chances of attaining practical and efficient solutions.

DOI: 10.1201/9781003542896-2

2.2 THE PROBLEM STATEMENT

We want to generate an efficient algorithm for the optimization problem presented in (2.1). Thus, it is necessary to consider three distinct spaces: algorithms (Ω_A), parameters settings for each algorithm a (Ω_{P_a}), and problem instances (Ω_I). We define a master optimization problem to search the three spaces simultaneously for the best possible algorithm. Figure 2.1 depicts the connection between the three spaces. Note that each algorithm has a parameter setting space. Thus, selecting one component of each space makes it possible to access the master problem's optimal solution.

$$\text{Minimize } f(x) \text{ subject to } x \in \Omega \tag{2.1}$$

Let us consider the following definitions:

- x: decision variables within the feasible space Ω for the problem (2.1).
- a: is an algorithm capable of finding a solution for the problem (2.1).
- p_a: is the parameter settings for the algorithm a. These parameter settings are in the space of possible parameter settings, Ω_{P_a}.
- i: denotes a problem instance in the space of possible problem instances Ω_I.

The master problem Π specified in Equations (2.2) and (2.3) has an objective function max $F(x,a,p_a,i)$ combining the components to guide the search of an algorithm for the problem (2.1). Such a function aims to maximize the algorithm's performance in solving the set of problem

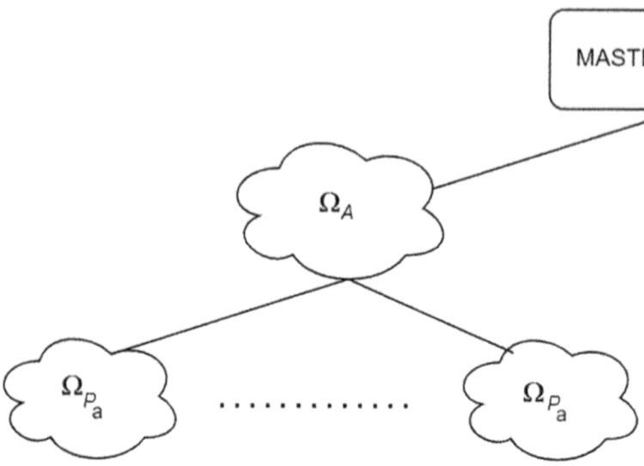

FIGURE 2.1 Schematic view of searching spaces of an optimization problem.

instances with a parameter setting. In turn, maximizing the performance means minimizing the error E incurred by the algorithm solving the set of instances. Measuring this error requires a reference solution for the optimization problem (2.1), denoted as $f^* = f(x^*)$. The specific form of this objective function would depend on the details of each optimization problem and the relationships between these variables.

$$\Pi: \text{ Min } E(x, a, p_a, i) \tag{2.2}$$

$$\text{Subject to:}$$

$$x \in \Omega,\ a \in \Omega_A,\ p_a \in \Omega_{P_a},\ i \in \Omega_I \tag{2.3}$$

Example 2.1

Formulate the master problem for the Weight Constrained Shortest Path Problem (WCSPP).

Problem definition:

In the WCSPP, the goal is to find a path from a source node s to a target node t in a graph such that the total cost of the path is minimized while adhering to a constraint on another metric, such as weight. Thus, given a graph $G = (V, E)$ with two metrics over paths, c (the cost) and w (the weight), and considering π as a path from s to t, and W' the predefined weight limit, the problem can be expressed as:

$$\min_{\pi} c(\pi) \quad \text{subject to} \quad w(\pi) \le W' \tag{2.4}$$

The master problem:

We want to generate an efficient algorithm for the optimization problem presented above. Thus, it is necessary to consider three distinct spaces: WCSPP algorithms (Ω_A), parameter settings for each WCSPP algorithm a (Ω_{P_a}), and WCSPP problem instances (Ω_I). We define a master optimization problem to search the three spaces simultaneously for the best possible algorithm.

Let us consider the following definitions:

- x: decision variables within the feasible space Ω for the WCSPP.
- a: a WCSPP algorithm capable of finding the optimal solution.
- p_a: the parameter settings for the algorithm a. These parameter settings are in the space of possible parameter settings, Ω_{P_a}.
- I: denotes a WCSPP instance.

The master problem Π is specified with an objective function $\min E(x,a,p_a,i)$ combining the components to guide the search for an algorithm. This function aims to minimize the error incurred by the algorithm when solving the set of instances. Consider the minimization of the mean squared error and introducing penalty functions to avoid algorithms with more than a predefined number of instructions, a predefined maximum running time, and a predefined maximum polynomial complexity, the master problem can be formulated as follows:

$$\Pi: \ \min E(x,a,p_a,i) = \frac{1}{|I|}\sum_{i\in\Omega_I}\left(c(\pi_i)-c_i^*\right)^2 + \lambda_1 P_{\text{instr}}(a) + \lambda_2 P_{\text{time}}(a) + \lambda_3 P_{\text{complexity}}(a) \tag{2.5}$$

Subject to:

$$x \in \Omega; \mathrm{a} \in \Omega_A; \ p_a \in \Omega_{P_a}; \ i \in \Omega_I; \ w(\pi) \leq W' \tag{2.6}$$

Where:

- $P_{\text{instr}}(a)$ is a penalty function for algorithms with more than a predefined number of instructions.
- $P_{\text{time}}(a)$ is a penalty function for algorithms exceeding a predefined maximum running time.
- $P_{\text{complexity}}(a)$ is a penalty function for algorithms exceeding a predefined maximum polynomial complexity.
- λ_1, λ_2, and λ_3 are penalty weights that balance the importance of minimizing the error and adhering to the constraints on instructions, running time, and complexity.
- $c(\pi)$ is the cost function of the path π.
- $w(\pi)$ is the weight function of the path π.
- c_i^* is the optimal cost, for problem instance i.

This formulation ensures that the generated algorithms are accurate, efficient, and practical for real-world use, addressing the constraints and objectives of the WCSPP.

2.3 THE SPACE OF INSTANCES

The space of instances of an optimization problem is composed of complexities that can strongly affect the performance of an algorithm. Each instance has a particular structure, and another instance may have a very different structure, which, therefore, offers another type of difficulty to the search process performed by the algorithm. This fact means that the same algorithm can have an excellent performance with one problem instance but a poor performance with another problem instance, which could even be very close. In other words, the space of instances has a nature that should be considered during the design of an algorithm. It has been a common practice that the design of optimization algorithms occurs without considering the space of instances, which eventually causes the algorithm to perform well with only a few instances. Furthermore, the performance may vary between two instances in terms of the quality of the solution or the computational time required to find it.

Identifying the role of problem instance features during the algorithm design is helpful. Problem instance features can be precious, although their role may not always directly predict algorithm performance for a given instance. While problem instance features may not directly predict algorithm performance for a given instance, they play a crucial role in advancing algorithm design and understanding the factors influencing algorithm behavior. They enable researchers and practitioners to decide which algorithms to consider for specific problems and provide valuable insights into the interplay between problem structure and algorithm performance.

Problem instance features can provide insights into the instances in which specific algorithms or algorithmic configurations tend to perform well. This knowledge is valuable both in the algorithm selection field and in the automatic generation of algorithms (AGA). Thus, this knowledge can guide the choice of algorithm for new instances based on similarities with instances where specific algorithms have historically performed well. Furthermore, problem instance features are valuable for understanding the relationship between problem structure and algorithm performance. Studying how different features correlate with algorithm behavior across diverse instances could help uncover patterns and insights. These insights can help design new algorithms or modify existing ones to handle specific structural characteristics more effectively. In this sense, problem instance features contribute to the broader algorithmic research and development field.

A good practice for studying problem instances is clustering them from their features. By analyzing the features of problem instances, one can group them into clusters based on similarities in their structural characteristics. Clustering algorithms like k-means, hierarchical clustering, or DBSCAN are commonly used for this purpose. Each cluster represents a subset of instances that share common traits. With clustered problem instances, designing or selecting algorithms tailored to perform well on each cluster is possible. These specialized algorithms are fine-tuned to exploit the specific characteristics and challenges posed by instances within a cluster. This task involves adjusting algorithm parameters, selecting appropriate algorithmic techniques, or even developing entirely new algorithms. The result is an algorithm for each cluster of problem instances. When faced with a new, previously unseen instance, its features allow us to determine which cluster it belongs to and select the corresponding specialized algorithm.

Example 2.2

Identify a set of traveling salesman instance features.

Consider the Euclidean TSP. The objective is to find the shortest route that visits each city of a set of cities exactly once and returns to the original city. The set of cities is located in the Euclidean plane, and the distance between two cities is the Euclidean distance. This TSP is considered the symmetric version because both distances outward and return are the same. Thus, the data for each instance is the number of cities, the coordinates for each city, and a matrix C, in which each element (i, j) contains the distance between both cities. Thus, C is an upper triangular matrix. Then, the following features can be used to classify the instances:

a) The number of cities.
b) The upper triangular distance matrix.
c) The longest to the shortest distance ratio.
d) The maximum number in each row of C.
e) The minimum number in each row of C.
f) The average distance in matrix C.
g) The number of city clusters.
h) The density of each city cluster.
i) Coordinates of cities.
j) Optimal tour length if available.

2.4 THE SPACE OF ALGORITHMS

The space of algorithms is composed of all possible algorithms. When we limit ourselves only to an optimization problem, the space is limited to all possible algorithms that solve specific problems. Contained in that space are exact algorithms that come from mathematical programming and algorithms that guarantee only an approximate solution to the problem. A problem approached via mathematical programming has one or more algorithms directly associated. For example, if the problem is approached via linear programming, the solution can be found with the simplex or interior point algorithms. When only an approximate solution to the problem is required, approximation heuristics, metaheuristic, or hyperheuristic algorithms are available. In addition, in the automatic algorithm generation approach, one can build a dedicated algorithm for the problem, a robust algorithm for a group of problem instances, or simply a dedicated algorithm for each problem instance. In the latter case, the algorithm is discardable because it is no longer useful once it has solved the instance.

Automating algorithm design and developing problem-specific or instance-specific algorithms can be seen as a second-generation approach to optimization problem resolution. In the first generation of optimization problem resolution, the focus was primarily on developing general-purpose algorithms that could be applied to various problem instances. These algorithms, often based on well-established mathematical programming techniques or heuristics, aim to provide near-optimal solutions in a general sense but might not take advantage of specific problem characteristics. With the advent of automation and the growing understanding of the importance of problem-specific or instance-specific approaches, a second generation emerged. In this generation, the emphasis has shifted toward creating algorithms tailored to the unique characteristics of individual problem instances or groups of instances. This approach involves automatically generating, adapting, or selecting algorithms based on the specific problem at hand. This second-generation approach represents a more sophisticated and tailored way of addressing complex problems. It leverages automation, data-driven techniques, and problem-specific insights to develop and apply algorithms better suited to each problem instance's characteristics, leading to improved optimization outcomes.

Several strategies can be employed to search in the space of algorithms, depending on the specific problem and goals. Initially, we can think of constructing algorithms from a set of building blocks or starting with an initial algorithm and exploring the algorithmic space from there. Furthermore, it is possible to use a metaheuristic to explore the space. Simulated annealing, or Particle Swarm Optimization, can be applied to visit the algorithmic space in a constructive or a single-solution manner. The selection of algorithm components or parameters can be treated as decision variables, and metaheuristics can be used to find optimal or near-optimal combinations.

Also, evolutionary algorithms, such as Genetic Algorithms, can be used to evolve algorithms over generations. This approach starts with a population of algorithms (represented as individuals), and through processes like mutation, crossover, and selection, different algorithms are evolved, improving their performance; that is precisely what genetic programming does. GP uses a syntactic tree as a representation on which an algorithm can be trivially mapped. Genetic programming then visits a sequence of algorithms guided by a fitness function, which, in this case, is defined as a combination of the efficiency of the algorithm and the value of the objective function of the optimization problem at hand.

The choice of strategy depends on the problem's nature, the algorithmic space's complexity, available computational resources, and the specific goals of the optimization process. Combining multiple strategies, such as using metaheuristics for algorithm configuration or incorporating reinforcement learning into algorithm selection, can also be effective in finding optimal or near-optimal solutions.

2.5 THE SPACE OF PARAMETERS

The parameter space comprises all possible numerical values for each parameter. Thus, searching this space for the best combination or setting is necessary to produce the best algorithm performance. Thus, tuning the parameters of algorithms is crucial in enhancing their performance and reliability. Every algorithm has default settings, which may not always be the most efficient for every unique problem or user requirement. By fine-tuning these parameters, it's possible to discover more suitable settings that boost the algorithm's effectiveness. This tuning process is not just about optimization; it also offers valuable insights into the relationship between problem instances, parameter values, and the algorithm's overall performance. This deeper understanding gained from tuning aids in calibrating and analyzing the algorithm's functionality.

The impact of parameter values on an algorithm's performance can vary greatly, with some parameters playing more pivotal roles than others. Taking the example of a genetic algorithm, aspects like population size, mutation rate, and crossover rate are often key parameters that substantially influence the algorithm's outcome. However, the significance of these parameters can differ based on the specific problem at hand and the characteristics of the genetic algorithm. This variability underscores the importance of tuning: it assists in pinpointing the most crucial parameters for a particular problem and helps determine their optimal values. Fine-tuning is thus essential for tailoring algorithms to specific needs and contexts.

2.6 SIMULTANEOUS PARAMETER OPTIMIZATION WITH AGA

Searching for optimal parameter settings for an algorithm is called the algorithm configuration, the hyperparameter optimization, or the tuning of parameters. This problem has been approached in various ways, either simultaneously with finding the optimal solution or as an independent problem. Setting parameters simultaneously with the search for the optimal solution for a specific set of problem instances can be conducted within a single optimization process of the optimization problem presented in Equations (2.7) and (2.8). So, metaheuristics like genetic algorithms or particle swarm optimization can handle both sets of decision variables.

$$\Pi_p\text{:}\quad \text{Min}\, F\left(x, a, p_a\right) \tag{2.7}$$

$$\text{Subject to}:$$

$$x \in \Omega, a \in \Omega_A, p_a \in \Omega_{P_a} \tag{2.8}$$

Let us consider as an example a genetic algorithm to solve Π_p. It is a task that requires several components. Given a feasible solution (x, a, p_a), the genetic operators and the GA parameters operators should be defined in such a way that when applied to the current solution, they produce a new Π_p solution. In this way, the GA string represents the candidate algorithm and its parameter settings. It consists of a combination of genes, where each gene encodes the algorithm and its parameters. In turn, the structure of the GA string depends on the granularity of control over the algorithm's design and parameterization. If we have a binary optimization problem, we can use 0–1 genes indicating the variable values. In addition, a similar representation identifies the presence or absence of a particular algorithmic component and possible parameter value. With this representation, a standard GA can be used in which selection, crossover, and mutation are well-known operators in evolutionary computing. Thus, the algorithm is decoded and run from this composed binary string, updating the current solution x.

The fitness function assesses how well a candidate solution performs on the optimization problem. Considering the algorithm's behavior and parameter settings, it should capture the objective to optimize. The fitness function guides the evolutionary process by assigning higher fitness values to better-performing solutions. In practice, developing a GA for this task is highly problem-specific, and the choice of GA string representation, genetic operators, and parameters depends on the nature of the optimization problem and the desired level of control over the generated algorithms. The GA hyperparameter tuning is also essential to ensure it performs well across various problem instances.

Example 2.2

Propose a simulated annealing solution representation to search for an algorithm and simultaneously search for the parameter setting to find solutions for the maximum clique problem in a graph.

The maximum clique problem is a graph theory problem that involves finding a subset of vertices in a given graph, where an edge directly connects every pair of vertices in the subset, and this subset is as large as possible. In other words, a subgraph (a subset of vertices) is considered a maximal clique if it is a complete subgraph (a clique), and there is no way to include an additional vertex from the original graph in this subset without breaking the completeness condition. The goal is to identify the largest such clique within the graph.

To approach problem (2.7)–(2.8) by SA, it is necessary to define a solution representation, a neighborhood generation, and an objective function.

- Solution Representation. Each SA solution is a tuple (x, a, p), where x represents the vertices forming a feasible maximum clique on the graph, a represents an algorithm or algorithmic components, and p is the algorithm's parameter settings.
- Neighborhood Generation. In each SA iteration, a neighboring solution is generated by randomly perturbing the current solution $s = (x, a, p)$ to obtain a new solution $s' = (x', a', p')$ that contains different algorithmic choices, parameter settings, and possible modifications to the current clique.
- Objective Function. This function evaluates the quality of the feasible maximum clique solution obtained using algorithm a with parameter settings p on the given graph. Additionally, the cost function can consider penalties or rewards for the quality and size of the clique solution.

2.7 INDEPENDENT PARAMETER OPTIMIZATION

A second approach is to treat the problem as an independent parameter optimization problem. The search for optimal parameter settings is a separate optimization problem. This approach allows for fine-grained control over the parameter optimization process, potentially leading to better-tuned algorithms and keeping the algorithmic search for solutions focused solely on finding the best solution, simplifying the optimization landscape.

$$\Pi_p^a: \ \text{Min}\, F(x, p_a) \tag{2.9}$$

$$\text{Subject to:}$$

$$x \in \Omega,\ p_a \in \Omega_{P_a} \tag{2.10}$$

2.8 THE ALGORITHM SELECTION PROBLEM

Given the vast array of potential algorithms available—including exact methods, heuristics, and metaheuristics—the task of selecting the most effective algorithm for a specific NP-hard problem becomes critical. This problem has been considered in the literature as the Algorithm Selection Problem (ASP). The core challenge is to develop a systematic and reliable method for selecting the most appropriate algorithm for solving a given instance of an NP-hard combinatorial optimization problem. This selection should optimize for solution accuracy, computational efficiency, and resource utilization. Developing an effective algorithm selection system could significantly enhance the capability to solve NP-hard combinatorial optimization problems more efficiently. This system will reduce computational costs and improve solution quality, thereby substantially impacting fields where these problems are prevalent.

The master problem is also a framework to represent the ASP. Given a set of k known algorithms $\Omega_A^s = \{a_1, a_2, \ldots, a_k\}$, to select the best algorithm for a given problem instance, the master problem becomes equation (2.11) and (2.12).

$$\Pi^s: \ \text{Min}\, F(\text{x}, a, p_a) \tag{2.11}$$

$$\text{Subject to:}$$

$$x \in \Omega,\ a \in \Omega_A^s,\ p_a \in \Omega_{P_a} \tag{2.12}$$

2.9 THE NO-FREE-LUNCH IN AGA

The No Free Lunch (NFL) theorem provides a fundamental insight that is especially relevant when considering the AGA master problem formulation in NP-hard combinatorial optimization problems. The NFL theorem states that no algorithm can outperform all others across all possible problem instances. This fact implies that an algorithm's performance is inherently problem-dependent; an algorithm that excels in one type of problem instance may perform poorly in another. In the context of the master problem, this theorem suggests the necessity of a tailored approach to the AGA. The master problem's objective, which is to minimize the error or maximize the performance of an algorithm for a given set of problem instances, aligns perfectly with the NFL theorem's premise. Thus, the optimal choice of an algorithm and its parameters is highly contingent on the specific characteristics of the problem instance and the nature of the optimization problem.

In addressing the master problem, one must navigate the diverse landscape of algorithms and their parameter settings, acknowledging that each combination may yield varying levels of efficacy depending on the problem at hand. This challenge is amplified in NP-hard combinatorial optimization problems, where the complexity and diversity of instances are significant. The master problem formulation, focused on systematically finding the most appropriate algorithm and parameter setting for each instance, becomes a direct response to the NFL theorem's implications. It necessitates a deep understanding of the problem feasible region and the algorithm space, as well as the intricate interplay between the algorithm's parameters and the instances. The objective function F, aimed at minimizing error or maximizing performance, must be dynamically adaptable and sensitive to the nuances of each problem instance.

2.10 SUMMARY

This chapter provides an in-depth exploration of the master problem in the context of optimization, focusing on selecting the most efficient algorithm for solving specific optimization problems. It discusses the complexities of searching the spaces of algorithms, parameters, and problem instances to maximize algorithm performance while minimizing new algorithm errors. The chapter delves into the inherent challenge posed by NP-hard problems due to the diversity of problem instances and the necessity for a tailored approach in algorithm selection. It aligns with the NFL theorem, suggesting that the effectiveness of an algorithm is contingent upon the specific characteristics of the problem instance, a fact that requires a comprehensive understanding of the interplay between algorithms, their parameters, and the problem space to systematically identify the optimal algorithm component combinations for each unique instance.

EXERCISES

1. Describe the master problem in your own words, including its objective function and constraints. Explain how the spaces of algorithms, parameter settings, and problem instances interact in the master problem. Provide a diagram similar to Figure 2.1 to illustrate these interactions and discuss the challenges of searching these spaces simultaneously.

2. Formulate the master problem's objective function for automatically generating algorithms that solve the graph coloring problem. This function should minimize the error between the number of colors the generated algorithm uses and a known optimal solution for a set of training graph instances. However, it should also balance this with factors like minimizing color conflicts (solution quality) and the computational efficiency of the generated algorithms across various graph instances.

3. Formulate the master problem's objective function to automatically generate algorithms for the multidimensional knapsack problem (MKP) by using simulated annealing to search the space of MKP algorithms. Inspired by metal cooling, SA overcomes local optima by initially allowing the exploration of worse solutions. The MKP involves selecting items with value but consuming resources across multiple dimensions. The goal is to maximize value within constraints. This formulation should identify (a) a solution representation, (b) a method for generating new candidate algorithms, and (c) a temperature decay schedule.

4. Consider the MKP described in exercise 2 to identify relevant features to cluster MKP instances.

5. Define the space of problem instances for the TSP, including relevant features that can influence the performance of algorithms. Discuss how these features can be used to cluster problem instances and guide the design or selection of algorithms.

6. Quantify the space of algorithms for the Binary Knapsack Problem (BKP) when the algorithms are represented as trees with functions in internal nodes and terminals in leaf nodes. First, a solution container will be defined to store the current solution found by a BKP algorithm. Terminals should be actions that act on the solution container. Functions in internal nodes like AND, OR, and WHILE will work by combining the operations of the BKP algorithm produced during the solution of the master problem. Begin by describing two known algorithms that solve the BKP. Explain how each algorithm can be represented in the context of the master problem using a tree structure. Thus, estimate all the possibilities of completing trees with a given depth.

7. The Vehicle Routing Problem (VRP) involves finding efficient delivery routes for a fleet of vehicles with capacity constraints, minimizing total distance while visiting each location exactly once. Iterated Local Search is a powerful technique that escapes local optima by combining local search improvements with solution disruptions. Formulate the master problem's objective function to generate VRP-solving algorithms using ILS automatically. This function should identify (a) a solution representation, (b) a local search operator, and (c) a perturbation strategy for diversifying solutions.

8. The Facility Location Problem involves finding optimal locations to place facilities to minimize total transportation costs. Tabu search is a metaheuristic that avoids revisiting recently explored solutions to escape local optima. Formulate the master problem's objective function to generate algorithms for the Facility Location Problem using Tabu Search. This function should specify: (a) a solution representation, (b) a neighborhood generation method for creating new candidate solutions (e.g., swapping tree nodes), and (c) a tabu list management strategy to prevent revisiting similar solutions too soon.

CHAPTER 3

Modeling Problems

In Operations Research (OR), Analytics, and Artificial Intelligence (AI), approaching a practical problem starts with generating a model of the situation. This modeling process allows for detailed study and analysis without intervening in the real problem, making it a crucial step in problem-solving. This chapter provides an overview of the modeling process, highlighting its importance in OR and AI, and explains how to model with AGA. Modeling provides a structured way to translate real-world issues into mathematical formulations or computational models. Practitioners can devise efficient, cost-effective, and viable solutions to complex operational problems by understanding and creating models. The goal is to equip the reader with the knowledge and tools needed to model real-world problems effectively, paving the way for innovative algorithmic solutions through AGA.

3.1 THE MODELING PROCESS

Transitioning from a well-defined real-world problem to its resolution necessitates a crucial intermediate phase: formulating a model that encapsulates the problem's fundamental attributes. This model functions as a substantive or graphical abstraction of the issue, serving as an interface between conceptual challenges and pragmatic solutions. The principal objective of this model transcends mere representation; it facilitates the manipulation and interaction with the problem domain in a significantly more tangible and perceptible manner for human operators. Such model-based interactions may manifest in physical form within

DOI: 10.1201/9781003542896-3

controlled experimental environments or in abstract representations on pedagogical surfaces and computational simulations, embodying the essence of the modeling process.

Models function as critical abstraction mechanisms wherein real-world phenomena' multidimensional, often stochastic complexities are reduced to more tractable, analyzable forms, facilitating rigorous exploration and manipulation beyond the constraints of the actual system. A simulation domain is generated by embedding an isomorphic component within the model that emulates real-world dynamics. While abstracted, this domain preserves the salient characteristics of the problem, enabling practitioners to conduct scenario analyses and system-level explorations with a degree of flexibility rarely afforded by empirical constraints.

In information technology, models emerge as crucial constructs, frequently articulated through the formalism of mathematical notation. Consider the optimization of resource allocation within an enterprise engaged in the production of a specific commodity. In this context, formulating a mathematical model necessitates establishing a system of equations to delineate the feasible region, creating a bounded space within which the optimal solution must reside. This process involves the identification of a feasible solution set. Furthermore, a mathematical objective function is defined, articulating a quantifiable metric to evaluate the solution quality, characterized by either the maximization or minimization of a particular measure.

Delving further, we enter the domain of mathematical programming models, which facilitate the precise representation and modeling of an extensive array of real-world scenarios. This comprehensive class of models, characterized by diverse methodologies and applications, enables practitioners to accurately simulate real-world situations, providing a structured framework for solution development and evaluation. Mathematical programming, encompassing a robust spectrum of models including linear, integer, and nonlinear programming, permits the simulation and exploration of diverse situational outcomes, each governed by its specific constraints and objectives.

However, the pursuit does not culminate with the modeling. The subsequent juncture in this journey involves identifying the most promising algorithm that seeks the optimal solution within the predefined mathematical framework. Determining the optimal algorithm is a nuanced task, considering not only the problem and model's specificities but also computational efficiency and practicality. Once the algorithm is identified or developed, it becomes the driving force behind the final computational tool. It is designed to be user-friendly and practical for the end-users.

This methodology offers a structured, systematic pathway from initial problem identification to mathematical modeling, culminating in algorithmic determination and computational tool development. It leads from an abstract problem to a tangible solution, ensuring that the final tool, crafted to aid decision-making, is theoretically sound and practically applicable within the specified real-world context. This synergy between mathematical modeling, algorithmic application, and IT presents a robust approach to solving complex, resource-sensitive problems in myriad real-world scenarios.

Figure 3.1 represents a sequential process transforming a real-world engineering problem into a computational solution. It encapsulates the journey from conceptualization to practical application. On the left, a real-world situation is represented. The next step corresponds to the mathematical model, depicted by the various optimization and function-based equations. These formulas represent the abstraction of the real-world problem into mathematical terms. An algorithm obtains the solution for the mathematical representation. This step is critical as it involves translating the mathematical model into a procedure or set of rules a computer can execute. Finally, the "Computer system" represents the implementation phase, where the algorithm is embedded into a software tool that allows the problem to be analyzed, predictions made, or outcomes optimized.

The scheme to model also implicitly describes the relationship between science and engineering. When the real-world problem lies on a computationally difficult problem, science takes place. The best way to solve the theoretical problem is to explore them scientifically. In this context, the synergy between science and engineering demonstrates how theoretical and practical realms interact. Science delves into the understanding and theoretical solution of complex problems, while engineering translates

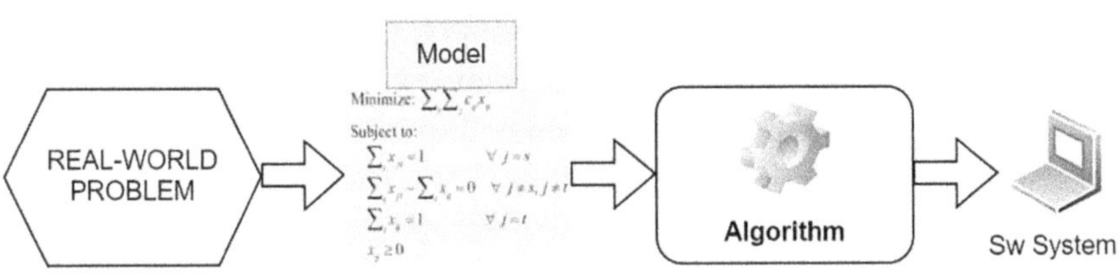

FIGURE 3.1 Schematic view of modeling.

these solutions into practical, usable computer systems. Together, they form a comprehensive approach to problem-solving, bridging the gap between abstract concepts and real-world applications. This collaboration is fundamental in tackling some of the most challenging problems faced in various fields, from technology and industry to environmental and social issues.

3.2 IDENTIFYING PROBLEMS

A solution is necessary when we identify that a real-world situation is a problem. However, identifying such a situation is also a problem. Fortunately, there are many approaches to characterize a problem. This task often involves recognizing a gap between the current state of things and a desired state and the existence of obstacles or challenges that prevent the transition from one state to another. Indicators represent the existence, status, or potential emergence of a problem in real-world situations. They provide a lens to recognize issues, allowing stakeholders to identify and address challenges before they escalate or persist. Indicators might manifest as physical symptoms, statistical outliers, behavioral changes, or any detectable sign that points toward an underlying issue. For instance, in a business environment, indicators like declining sales, reduced customer satisfaction, or increased employee turnover might signal problems in strategy, product quality, or organizational culture.

These indicators can be qualitative or quantitative and are crucial in facilitating early intervention, enabling proactive rather than reactive responses. Qualitative indicators might include subjective experiences, such as stakeholder dissatisfaction or emerging negative perceptions in a community. On the other hand, quantitative indicators utilize numerical data, such as performance metrics or incident rates, to highlight potential problems. Attuning to such indicators allows for timely problem identification, enabling decision-makers to initiate problem-solving processes and implement corrective actions, thereby bridging the gap between the existing and desired states. This attunement safeguards systems, organizations, and communities' health, sustainability, and progression.

Three steps can help to identify problems: observation and awareness, analysis and evaluation, validation and definition:

- Observation and awareness. The initial phase in identifying a problem necessitates a vigilant observation and an acute awareness of the environment and processes within it. This step implies noticing changes, irregularities, or disparities in the anticipated versus actual outcomes. Observing can encompass monitoring elements like performance metrics, behavioral patterns, and environmental conditions. Gathering feedback from stakeholders, such as employees, customers, or community members, is vital in identifying potential issues. Their experiences and insights often spotlight areas that may otherwise go unnoticed, unveiling hidden inefficiencies, dissatisfaction, or arising needs that warrant attention.

- Analysis and evaluation. Once observations and preliminary information are available, detailed analysis and evaluation become pivotal to further identifying potential problems. Employ data analytics to scrutinize numerical information for patterns, anomalies, or deviations that might hint at underlying issues. Here, it is crucial to assess and compare the existing situation against established benchmarks, norms, or similar scenarios to pinpoint discrepancies. Evaluating feedback and information through various lenses, such as socio-economic, environmental, or technological perspectives, aids in comprehensively understanding the facets and repercussions of the identified issues. An integrated analysis combining quantitative and qualitative data offers a well-rounded perspective on potential problems.
- Validation and definition. The validation third step ensures that the identified issues are not sporadic or isolated events but genuine, persistent problems that require resolution. This step involves checking the consistency of the problem and its impacts across various parameters and scenarios. Once a problem is validated, crafting a succinct definition becomes paramount. This definition should encapsulate the issue's essence, elucidating who is affected, what the adverse effects are, where it is occurring, and, whenever possible, hint at why it is happening. Ensuring that the problem is defined with clarity and precision sets the foundation for subsequent problem-solving endeavors, paving the way for designing effective solutions and strategies.

A well-constructed problem statement can guide subsequent problem-solving processes, focusing on developing solutions and strategies. A systematic approach to identifying and defining problems helps ensure the issues are well-understood and addressed effectively. In this sense, we can define a checklist that, although it does not cover the universe of all problems, works as a guide to identify a situation that needs attention because it is a problem. We propose an eight-item checklist.

a) Gap Measurement: Is there a discernible gap between the actual and desired outcomes?

b) Obstacle Identification: Are there barriers preventing the achievement of objectives?

c) Performance and Standards: Are there deviations from established performance benchmarks or standards?

d) Stakeholder Concerns: Have stakeholders raised any concerns, complaints, or negative feedback?

e) Negative Consequences: Are negative impacts, risks, or potential losses associated with the current situation?

f) Legal and Ethical Compliance: Does the situation comply with legal regulations and ethical standards?

g) Resource Evaluation: Are resources being used inefficiently, and is the situation sustainable in the long term?

h) Data-Driven Indicators: Do the data and metrics indicate significant anomalies or trends that suggest a problem?

This list serves as a diagnostic tool rather than a strict scoring system. All questions don't need to be answered affirmatively to identify a situation as a problem. Often, a single critical issue can constitute a severe problem. However, the number of affirmative or partially affirmative responses can help gauge the severity and complexity of the problem. The more questions that receive a "yes" or "partially" response, the more multifaceted and potentially urgent the problem may be. In practice, we can consider that one or two affirmative responses may indicate a specific issue that requires targeted attention. Several affirmative responses suggest a broader and more complex problem that may need a comprehensive approach. Most of all affirmative responses reflect a critical, systemic problem that could require immediate and extensive intervention. Additionally, there is a relationship to be analyzed between the items with affirmative or partially affirmative responses. For instance, a problem that impacts critical performance indicators and violates legal standards (even if it's just two affirmative responses) could be far more pressing than several lesser issues that don't affect core operations or compliance.

3.3 APPROACHING PRACTICAL PROBLEMS IN OPERATIONS RESEARCH

Modeling is indisputably a pivotal step in Operations Research (OR. It provides a structured and systematic approach to analyzing real-world situations and making informed decisions. It offers a simplified problem representation, making it amenable to analytical and computational methods. In OR, the modeling process involves translating a real-world problem with all its intricacies into a mathematical formulation or a computational model.

In OR, creating models helps dissect problems related to resource allocation, logistical planning, operational efficiency, and many other decision-making dilemmas. By employing various mathematical and computational models, such as linear programming, network flows, machine learning, and simulation models, practitioners can scrutinize different scenarios, evaluate various strategies, and predict the impacts of different decision alternatives.

The model serves as a magnifying glass, enabling researchers and professionals to focus on the critical aspects of a more complex problem while temporarily sidelining the lesser impact factors. Furthermore, these models facilitate testing the potential solutions in a risk-free virtual environment before any real-world implementation, thereby saving resources and mitigating potential negative implications. Hence, in the context of OR, modeling isn't just crucial; it's integral, aiding in devising efficient, cost-effective, and viable solutions for intricate operational problems.

A systematic and methodical approach in OR is used to construct a problem model. The methodology is crucial to tackling the complexities and subtleties of real-world problems, ensuring that the developed models represent the issues at hand and are robust and reliable in deriving meaningful, applicable solutions. The systematic approach typically involves stages like understanding the problem, gathering and preprocessing relevant data, formulating a mathematical or computational model, validating and testing this model, solving it using appropriate algorithms, and implementing and validating the derived solutions in the real-world scenario. Further, the model's continuous monitoring and iterative refinement are also involved to ensure its sustained relevance and efficacy.

This structured approach is pivotal in bridging the gap between theoretical modeling and practical applicability, ensuring that the solutions derived are theoretically optimal, practically implementable, and impactful. It allows for coherently translating often intricate problems into structured models. This systematic strategy ensures that every aspect of the problem is considered, enhancing the likelihood of success when solutions are implemented in real-world scenarios. The following steps are functional to approach a real-world situation with an OR model:

- **Step 1. Understanding the Problem Domain.** The initial phase in problem modeling entails a comprehensive analysis of the problem domain, involving immersive investigation to identify and delineate the problem's scope, emphasizing macroscopic and microscopic contextual factors. Concurrently, it is imperative to crystallize the model's objective function, ensuring subsequent steps are unambiguous and directed toward a quantifiable target. In parallel, researchers implement a rigorous data acquisition and preprocessing protocol. This protocol encompasses a systematic collection of pertinent data, including variables and parameters crucial to the problem's multidimensional context, followed by thorough data validation to ensure reliability and consistency and subsequent preprocessing and transformation to prepare the data for model development.

  Moreover, a critical activity in this phase is conducting a literature review to identify similar problems studied in the literature or approached with existing software tools. This review offers valuable insights into previously used methodologies, helps us avoid redundant efforts, and can suggest potential solutions and tools that we may use to adapt to the current problem.

  This phase culminates in the generation of a formal problem statement document, accompanied by a curated dataset, serving as a foundational artifact that encapsulates a precise problem definition, delineation of contextual boundaries, specification of the objective function, and an inventory of available data, including metadata on its provenance, quality, and preprocessing methodologies applied. Additionally, the document should include findings from the literature review, highlighting relevant studies and existing tools. This comprehensive approach forms the basis for the subsequent model formulation and parametrization stages.

- **Step 2. Formulating the Model**. The model is mathematically formulated by identifying pertinent variables, establishing one or several objective functions, and defining constraints. The model strategically intertwines these elements, creating a mathematical or computational representation that mirrors the real-world problem. This synthesis captures the objective and the limitations (constraints) within a structured expression, which subsequently serves as a navigational blueprint toward seeking optimal solutions.
- **Step 3. Selecting or Designing an Algorithm**. An algorithm is selected to solve the model. This task involves selecting a suitable algorithm to find the best solution for several scenarios. In some situations, it could be necessary to design an algorithm for the model manually or automatically. During this phase, we must make pivotal choices about the programming language or environment for implementing the algorithm computationally. One may design the algorithm from scratch or leverage an existing algorithm. For instance, in Linear Programming, a plethora of robust software suites is equipped with an array of pre-built algorithms pertinent to the domain.

  A cost-benefit analysis considering the full scope of the project should be undertaken to guide these choices, weighing factors such as development time, performance, ease of use, and overall financial implications to inform the decision-making process. The model is gradually refined by computationally testing it. Ensuring that the model mirrors the real-world scenario faithfully and adheres to initial assumptions is crucial to circumvent potential pitfalls. Testing the algorithm with various data sets, parameters, and scenarios ensures its reliability and robustness, facilitating a thorough evaluation before any potential implementation in a practical context. The output is a computer code documented with the experimental analysis.
- **Step 4. Computer Implementation of the Algorithm**. The fourth step transitions the problem-solving process from theoretical models to practical applications by embedding the selected or newly designed algorithm computer code within a computer system. This crucial phase encompasses the development of software capable of executing the algorithm efficiently and reliably. Software Engineering principles actively guide the creation of user-friendly software. They enable users to input data directly or from other computer systems,

run algorithms, interpret results, and store information effectively. The application must undergo rigorous testing during a start-up period because it is necessary to validate that the computer system performs accurately under various operating conditions and that the software is robust to data anomalies and user errors in the working environment in which the problem occurs.

This stage often includes debugging, optimization for performance, and scalability assessments to ensure the tool can handle the anticipated load and data volume. The successful execution of this step culminates in a deployable software solution poised to tackle the original real-world problem with precision and adaptability, bridging the gap between the theory and tangible, actionable outcomes. The final output is a computer system that executes the required computations and is integrated into the existing technological ecosystem, ready for real-world deployment.

- **Step 5. Continuous Improvement**. The OR model and the software are not static entities but dynamic tools, subject to periodic reviews, revisions, and refinements. It embodies a commitment to perpetual improvement, ensuring the model remains relevant, responsive, and retentive of its efficacy in evolving real-world contexts. Insights from ongoing reviews are harnessed to refine and recalibrate the model, ensuring it continually aligns with and adeptly addresses the complexities and changes in the problem domain.

3.4 FUNDAMENTAL MODELS IN OR

Precast models in mathematical programming, exemplified by classic problems like linear programming, the traveling salesman, or the knapsack problem, serve as foundational templates for modeling a wide range of real-world situations. These models are akin to blueprints: they provide a starting point for structuring complex problems mathematically coherent and solvable. For instance, the traveling salesman problem at its core deals with finding the most efficient route, a concept applicable to logistics, delivery systems, and circuit design. Similarly, the knapsack problem's principle of optimizing the combination of items under a constraint finds relevance in resource allocation, budget management, and portfolio optimization. These precast models come with a well-established theoretical base and tested solution methods, offering a significant advantage in modeling. They enable practitioners to quickly map real-world scenarios onto these frameworks, adapting and extending them as necessary. This approach accelerates the problem-solving process and ensures rigor and reliability, as these models have been extensively studied and validated in various contexts. Hence, these precast models in mathematical programming are invaluable for efficiently transforming real-world complexities into manageable and solvable mathematical models.

The following two classical mathematical models provide an excellent example of fundamental structures that can be solved using various optimization algorithms. More detailed formulations may be required depending on the specific problem instance and additional constraints. However, such classical models can significantly help understanding some real-world situations. Although these models are relevant, many others are available in the literature, covering many natural situations.

a) Traveling Salesman Problem. Given a set of cities and the distances between each pair of cities, find the shortest possible route that visits each city exactly once and returns to the origin city. Let d_{ij} be the distance between cities i and j, and x_{ij} be a binary variable equal to 1 if the route goes directly from i and j and 0 otherwise. Furthermore, let's introduce auxiliary continuous variables u_i for each city, representing the order in the tour. The model for this situation is:

$$\text{Minimize} \sum_{j \neq i = 1}^{n} d_{ij} x_{ij}$$

Subject to:

$$\sum_{i=1, i \neq j}^{n} x_{ij} = 1, \quad \forall j$$

$$\sum_{j=1, j \neq i}^{n} x_{ij} = 1, \quad \forall i$$

$$u_i - u_j + n x_{ij} \leq n - 1, \forall 1 \leq i \neq j \leq n$$

$$x_{ij} = 0 \, or \, 1, \forall i, j u_i \in \mathbb{R}$$

This model minimizes the total distance of the tour as established in the objective function. The first set of constraints ensures that every city is entered and exited only once. The sub-tour elimination constraints prevent the formation of sub-tours, ensuring a single tour that visits every city once and returns to the starting city. Thus, this mathematical formulation of the TSP captures the essence of the problem by modeling the requirements of visiting each city exactly once and minimizing the total travel distance while ensuring that the solution is a single continuous tour without isolated loops or subtours.

The TSP has numerous real-world applications, such as logistics for optimizing delivery routes, manufacturing for planning the movement of machines, and circuit design for minimizing the wiring length. Additional constraints or objectives might include time windows, multiple depots, or vehicle capacity limits, adapting the problem to specific practical needs.

b) Knapsack Problem: Given a set of items, each with a weight and a profit, determine the number of each item to include in a collection so that the total weight is less than or equal to a given capacity and the total profit is as large as possible. Let p_i be the profit and w_i the weight of item i, x_i be the number of instances of item i to include in the knapsack and W the knapsack capacity. The model for this situation is:

$$\text{Maximize} \sum_{j \neq i=1}^{n} p_i x_i$$

Subject to:

$$\sum_{i=1, i \neq j}^{n} w_i x_i \leq W$$

$$x_i = 0 \, or \, 1, \forall i$$

In this problem, the objective is to maximize the total profit while ensuring that the total weight of the items chosen does not exceed the capacity of the knapsack. This problem is a classic example in combinatorial optimization and decision-making, illustrating how constraints can limit the feasible choices in optimizing an objective. It has applications in resource allocation, budgeting, and many areas where a trade-off between profit and capacity needs to be optimized.

The Knapsack Problem can be seen in many practical scenarios, such as selecting investments in a portfolio with a limited budget, choosing projects to maximize benefits under resource constraints, and cargo loading to maximize value without exceeding weight limits. Additional constraints or objectives might include multiple knapsacks, fractional selections, or varying item availability, making the model adaptable to real-world requirements.

In many real-world situations related to both problems, it is difficult to imagine that an actual situation fits precisely in this model; however, these fundamental problems are a starting point in the critical phase of modeling the situation. The Traveling Salesman Problem and the Knapsack Problem serve as foundational models, but real-world applications often require adjustments or additional constraints to represent the problem accurately. Many other fundamental problems, such as the Vehicle Routing Problem, the Job Shop Scheduling Problem, the Minimum Spanning Tree Problem, and the Bin Packing Problem, have been extensively studied in the scientific literature. Each of these problems provides a framework for understanding and tackling a wide range of practical challenges, illustrating the versatility and importance of these classical models in the field of optimization. By starting with these well-established models, one can build upon them to create more tailored solutions that address real-world situations' specific nuances and constraints.

3.5 APPROACHING PRACTICAL PROBLEMS VIA AI

AI has introduced a paradigm shift in modeling real-world problems, mainly through machine learning methods. This approach diverges from traditional OR methods by utilizing historical data as the foundation upon which predictive models are built and refined. In this paradigm, algorithms are trained on vast datasets that capture the nuances and patterns of the problem environment, learning to make predictions or decisions that improve over time. This data-driven methodology enables the model to adapt as more data becomes available, providing dynamic and often more nuanced solutions to complex problems such as consumer behavior prediction, natural language processing, and autonomous systems. The AI paradigm complements and sometimes transcends traditional

OR techniques, especially in scenarios where predictive accuracy and adaptability to new information are paramount.

AI, Analytics, and Machine Learning (ML) are interconnected domains that collectively revolutionize data interpretation and decision-making processes. The boundaries between fields are becoming blurred because they complement each other and share the goal of extracting valuable insights from data. AI represents the broader concept of machines performing tasks that mimic human intelligence, encompassing reasoning, learning, and self-correction. Within AI, ML stands as a subset, focusing specifically on algorithms that enable machines to learn from and make predictions based on data, effectively improving their performance over time without being explicitly programmed for each task. While overlapping with AI in the use of data to inform decisions, analytics traditionally relies more on statistical methods. It is broader in scope, including descriptive and diagnostic analyses to interpret past performance and infer future trends. While ML is inherently predictive and often prescriptive, operating on the premise of data-driven learning and adaptability, Analytics encompasses a broader range of data examination, from understanding "what happened" to advising what actions to take. Though distinct in their focus and methodology, the three fields increasingly blend in practice, leveraging their strengths to provide comprehensive insights and automated, intelligent decision-making tools.

A fundamental commonality among AI, Analytics, and ML is their reliance on historical data from various phenomena to derive insights into real-world problems. Such shared characteristics form the bedrock upon which these disciplines operate, albeit with different focuses and methodologies. Thus, a different paradigm for solving practical problems arises. The following equivalent to the OR steps allow us to approach a real-world problem using most of the methodologies existing in those fields:

- **Step 1. Descriptive Analysis and Problem Definition**. Begin by clearly defining the problem statement and objectives. Then, leverage AI-powered literature review tools to systematically search and analyze relevant scientific publications, research papers, and technological reports. This AI-assisted process helps identify similar already researched problems, potential solution approaches, and available technological tools. The AI can help summarize critical findings, highlight emerging trends, and identify gaps in current research. Concurrently, perform an initial exploratory data analysis using

AI and analytics techniques to understand patterns, anomalies, and trends in the available historical data. The output is a comprehensive problem definition, a synthesized literature review, and an initial data exploration report, providing a solid foundation for further analysis.

- **Step 2. Data Preparation and Predictive Analysis.** Synthesize insights from the AI-assisted literature review and initial data exploration to construct a comprehensive real-world model, leveraging fundamental AI optimization paradigms. Classify the problem within standard frameworks (e.g., classification, regression, clustering, etc.) and select appropriate algorithms (e.g., support vector machines, gradient boosting, DBSCAN, etc.). Implement AI-driven data preparation, including automated cleaning, feature engineering, and enrichment from identified external sources. Develop advanced predictive models using state-of-the-art machine learning and deep learning techniques, incorporating transfer learning from analogous domains. Utilize AutoML platforms to explore and optimize model architectures within the chosen paradigm efficiently. Rigorously validate and fine-tune models through cross-validation and Bayesian optimization. The output is a robust, AI-powered predictive model grounded in established optimization frameworks. It is accompanied by comprehensive documentation of its methodology, performance metrics, and limitations, thus providing a solid foundation for subsequent analysis and decision-making processes.
- **Step 3. Prescriptive Analysis and Solution Development.** In the third step, prescriptive analysis enters the scene. Here, actionable recommendations and decisions are derived from the insights gained from descriptive and predictive analyses. The developed models might recommend courses of action and potentially optimize for desired outcomes, ensuring solutions are data-driven and aligned with the defined objectives. Software engineers can transform prescriptive models into user-friendly tools if software development is required (e.g., development of decision-support tools). The output involves clear action recommendations, decision-making guidelines, and a software tool embedded with analytical models if relevant.
- **Step 4. Implementation, Monitoring, and Continuous Enhancement.** Post-development, the analytics solutions or recommendations are

implemented in the real-world scenario. Continuous monitoring is paramount to practically tracking the efficacy and accuracy of the analytical models and solutions. By employing a feedback loop, learnings from the implementation phase are iteratively fed back into the system, refining and enhancing the models and solutions for ongoing relevancy and optimization. Thus, in this phase, the output is the initial implementation and an ongoing, dynamic process of analysis, refinement, and adaptation to ensure enduring efficacy and relevancy in the ever-evolving problem domain.

3.6 FUNDAMENTAL MODELS IN AI

Similar to OR, AI also has its set of fundamental models that serve as foundational tools for various applications. The most used models include regression, classification, and clustering, which are central to AI addressing different problems and data characteristics. Let us formulate those problems:

a) Regression model
Consider a situation where a dependent variable y is a function of several independent variables $(x_1, x_2, \ldots, x_n)$. This function can be linear or nonlinear, typically expressed as Equation (3.1). The crucial task is ascertaining the precise parameters that characterize this function. Let's denote these parameters as $(\theta_1, \theta_2, \ldots, \theta_n)$, one corresponding to each independent variable. In addition, we consider an additional parameter θ_0, to consider the noise in the obtention of actual data. For example, we can represent the relationship in a linear model as Equation (3.2).

$$y = f\left(x_1, x_2, \ldots, x_n\right) \tag{3.1}$$

$$y = \theta_0 + \theta_1 x_1 + \theta_2 x_2 + \ldots + \theta_n x_n \tag{3.2}$$

To identify the optimal values of these parameters, we construct an optimization problem that explores the parameter space, seeking to align the model with the empirical data optimally. The objective function minimizes the discrepancies between the predicted values of y and their actual observed counterparts. The optimization task leverages historical data consisting of N observations of the system's functioning, where each observation i is denoted as $(y_i, x_{1i}, x_{2i}, \ldots, x_{ni})$. An observation represents a single data point or instance in the dataset. For example, in a house price prediction model, an observation might consist of features like square footage, number of bedrooms, location, and the corresponding sale price (y_i). Each house in the dataset would constitute one observation, collectively forming the basis for training and validating the regression model. This dataset helps evaluate the objective function across diverse points in the parameter space, aiming to pinpoint those parameter configurations where the predictive outcomes of the function most closely mirror the actual, historically observed data. Thus, for each data point, i Equation (3.1) produces an estimate of y, which we denote as $\hat{y}_i$ in Equation (3.3).

$$\hat{y}_i = f\left(x_{1i}, x_{2i}, \ldots, x_{ni}\right) \tag{3.3}$$

This process transforms the theoretical model into a practical tool calibrated against real-world observations, enabling it to provide actionable insights and predictions based on the underlying patterns and relationships within the data. The goal is to minimize the sum of errors between the values $\hat{y}_i$ and the y actual values. For example, the error can occur as the squared differences between predicted and observed values, given rise to a mathematical formulation of a regression problem as presented in Equation (3.4)

$$\text{Minimize } F\left(\theta_1, \theta_2, \ldots, \theta_n\right) = \sum_{i=1}^{N}\left(y_i - \widehat{y_i}\right)^2 \tag{3.4}$$

This optimization problem, being unconstrained, is amenable to various solution methods, such as gradient descent for iterative approximation or more direct analytical approaches in the context of linear regression, the latter yielding a solution through what is known as the Normal Equation. This solution method furnishes estimates for the parameter vector, $(\theta_1, \theta_2, \ldots, \theta_n)$, that aligns optimally with the observed data following the Least Squares criterion. Consequently, this approach leads to a definitive mathematical expression that characterizes the observed phenomenon within the dataset.

However, this framework does not inherently limit the nature of the relationship between the dependent and independent variables; it can encompass a broad spectrum of relationships. Real-world applications of regression models often necessitate a more intricate structure, potentially incorporating transformations of the independent variables, the inclusion of interaction terms, and integrating penalty functions. These enhancements are particularly crucial in high-dimensional spaces to mitigate the risk of overfitting, ensuring that the model fits the data well and maintains its predictive power and generalizability.

Equation (3.1) may be nonlinear and potentially non-differentiable, necessitating advanced optimization techniques beyond traditional gradient-based methods. In such cases, metaheuristic algorithms become valuable tools for parameter estimation. These techniques can navigate complex, multimodal solution spaces without requiring function derivatives. Ensemble or hybrid approaches combining multiple optimization strategies can enhance solution quality and robustness. The choice of optimization technique depends on the specific characteristics of the function, the dimensionality of the parameter space, and the computational resources available.

b) Classification model

Consider a scenario where we register the operation of a system in several operational variables that can be measured and stored as numerical or even as categorical values. Each time the system captures values, it generates a data point. We want to categorize or classify the set of data points into discrete classes or categories based on several independent variables, $(x_1, x_2, \ldots, x_n)$. Defining the variable y as the class label, we have a numerical representation of each data point's class. In this setup, each data point has a specific y value, where y denotes the class number or category to which the data point belongs. This approach allows us to formulate a mathematical equation representing the relationship between the class labels and the independent variable values, as in Equation (3.5). Similarly to the regression problem, it is necessary to identify the parameters of the equation Θ.

$$y = f\left(x_1, x_2, \ldots, x_n\right) \tag{3.5}$$

The objective function of the optimization problem to determine the parameters of Equation (3.5) aims to guide the search for the values of these parameters that most effectively classify the data points. Thus, the optimization problem explores the parameter space. The optimization process uses historical data comprising N data points, where each data point i is denoted as $(y_i, x_{1i}, x_{2i}, \ldots, x_{ni})$, with y being the actual category of the observation. The objective function in this context often aims to minimize discrepancies between the model's predicted category and the actual category. For example, a standard objective function in binary classification is the cross-entropy function, which penalizes incorrect predictions, especially those made with high confidence. With each data point belonging to one of the several classes involved in the dataset, Equation (3.6) is typical for obtaining the parameters. In that equation y_{ik}, is a binary indicator (0 or 1) if class label k is the correct classification for the $i-$th data point and, p_{ik}, be the predicted probability of the $i-$th data point for class k. All classes are considered when taking the sum, and we calculate the penalty for each class label. We use the negative sign to ensure that the function always remains positive.

$$\text{Minimize } F(\Theta) = -\frac{1}{N}\sum_{i=1}^{N}\sum_{k=1}^{K} y_{ik}\log(p_{ik}) \tag{3.6}$$

Once the objective function in an optimization problem, such as those found in regression or classification models, has been formulated, the optimization field's subsequent process of finding the optimal values for the equation's parameters is well-established. Various nonlinear optimization algorithms can be employed, ranging from exact methods like gradient descent to heuristic methods like genetic algorithms or simulated annealing when the problem is too complex or large-scale for exact solutions. Exact methods follow a systematic approach to finding the optimal solution, often requiring certain mathematical conditions like differentiability. In contrast, heuristic methods provide approximate solutions that are typically quicker to compute and useful for exploring vast or complex search spaces where exact methods may be computationally impractical. The choice between exact and heuristic methods depends on the specific requirements of the problem, such as the size of the parameter space, the complexity of the objective function, and the desired balance between computational efficiency and solution accuracy.

Real-world applications of classification models are vast, including image recognition, spam detection, and medical diagnosis, and often require complex model structures. These may include nonlinear transformations of the independent variables, interaction terms, and applying regularization (penalties) techniques to prevent overfitting, particularly in scenarios involving many predictor variables. The result is a model that classifies data accurately and adapts to the complexities and nuances inherent in real-world data scenarios.

c) Clustering model

Data collection is fundamental to unraveling inherent dynamics in any complex system, whether an intricate industrial operation, a dynamic social network, or a diverse ecological ecosystem. The challenge, however, lies in effectively interpreting this data, particularly in identifying and classifying clusters that exhibit similar characteristics. One of the possibilities is to resort to automatic clustering. Let us consider a dataset $P = p_1, \ldots, p_n$ comprising N data points and an associated matrix X of dimensions $(N \times t)$, where each row vector $x \in X$ uniquely represents the $i-$th data object in P. In this matrix, each element $\propto_{ij}$ within the vector x_i denotes the $j-$th real-valued feature $(j = 1, \ldots, t)$ of the $i-$th data point $(i = 1, \ldots, n)$. The objective is to methodically categorize the set P into a collection of clusters $C = \{C_1, \ldots, C_k\}$, where the number of clusters k is not predetermined. This optimization process aims to maximize individual clusters' similarity and the dissimilarity between distinct clusters. Three fundamental properties ensure a robust clustering: (i) each cluster C_i must contain at least one data point, ensuring $C_i \neq \emptyset \forall i \in \{1, \ldots, k\}$; (ii) any two distinct clusters C_i and C_j must be mutually exclusive, fulfilling $C_i \cap C_j = \emptyset \ \forall i \neq j$ and $i, j \in \{1, \ldots, k\}$; and (iii) every data point in P must be assigned to one of the clusters, thereby satisfying $\bigcup_{i=1}^{k} C_i = P$. This structured approach to clustering categorizes the data effectively and ensures comprehensiveness and exclusivity within the cluster formation.

To formulate the optimization problem for this task, we introduce a silhouette index that combines the concepts of cohesion and separation among data points. The average Euclidean distance between two data objects is denoted as d_{ij}, as defined in Equation (3.7). Additionally, we define intra-cluster similarity within a given cluster, denoted as C_w, using Equation (3.8), and inter-cluster similarity using Equation (3.9). The silhouette index $s(x_i)$ for a specific data point x_i is then defined by Equation (3.10). The difference $b(x_i) - a(x_i)$ quantifies how much more separated point i is from other clusters compared to how tightly it's grouped with points in its own cluster. A positive difference indicates that the point is well-matched to its cluster, while a negative difference suggests it might be better suited to a neighboring cluster.

By maximizing the average silhouette index, as expressed in Equation (3.11), we obtain a clustering solution based on each data point's silhouette index. The optimal number of clusters is determined by the k value corresponding to the highest silhouette index.

$$d(x_i, C_t) = \frac{1}{|C_t|} \sum_{\forall x_j \in C_t} d_{ij} \tag{3.7}$$

$$a(x_i) = \frac{1}{|C_w| - 1} \sum_{\forall x_j \neq x_i, x_j \in C_w} d_{ij} \tag{3.8}$$

$$b(x_i) = \text{Min}\, d(x_i, C_t), C_t \neq C_w, C_t \in C \tag{3.9}$$

$$s(x_i) = \frac{b(x_i) - a(x_i)}{\text{Max}\, b(x_i), a(x_i)} \tag{3.10}$$

$$\text{Max}\, SI = \frac{1}{n} \sum_{i=1}^{n} s(x_i) \tag{3.11}$$

Addressing the automatic clustering problem hinges on maximizing cluster similarity. Upon establishing this framework, the path to an optimal solution becomes more straightforward, albeit still challenging, due to the complexity of the problem. This optimization can use various techniques, ranging from standard optimization algorithms to more specialized algorithms explicitly designed for clustering tasks. These techniques could include gradient-based methods, evolutionary algorithms, or even bespoke heuristic approaches tailored to the unique nature of clustering problems. Each method has its strengths and is selected based on the dataset's requirements and the clusters' desired characteristics, such as size, number, or degree of separation. The ultimate goal is to navigate the search space effectively to find a clustering configuration that maximizes intra-cluster similarity while maintaining clear distinctions between different clusters.

The use of fundamental AI and OR models to represent a wide range of real-world scenarios underscores a fundamental approach in engineering and data science: the abstraction of complex realities into manageable mathematical models. These fundamental models serve as archetypes, offering a structured approach to problem-solving by quantifying and analyzing various aspects of real-world phenomena. Indeed, the first critical step in modeling is assessing whether a real-world scenario fits into an existing fundamental problem or requires a particular mathematical model. This task involves carefully examining the scenario's characteristics to determine if they align with the assumptions and structures of known models like those used in regression, classification, or clustering. However, it's common to find scenarios that do not neatly fit into these predefined molds.

In scenarios where direct application of fundamental models is impractical, their underlying principles serve as conceptual templates, guiding the development of bespoke solutions. This approach involves adapting, combining, or innovating modeling techniques to address specific problem nuances. Analogous to constructing a problem-solving toolkit, the study of fundamental models equips engineers with adaptable strategies. By leveraging this knowledge base, practitioners can extrapolate, modify, and synthesize novel approaches tailored to unique challenges. This methodology facilitates a more profound comprehension of complex problems and catalyzes innovation in model creation. Consequently, it yields more effective and efficient solutions that closely align with the intricacies of real-world scenarios, thereby enhancing the overall problem-solving efficacy and expanding the boundaries of applied modeling techniques.

3.7 APPROACHING PRACTICAL PROBLEMS BY AGA

Automatic generation of algorithms can enable OR models to adapt to dynamic problem environments. As conditions change, the system can automatically generate a new algorithm tailored to the problem, ensuring a near-optimal solution. Furthermore, complex OR problems often have multifaceted constraints and objectives. AGA can help identify and generate highly specialized algorithms that might be difficult or time-consuming for human researchers to develop, thereby streamlining the problem-solving process. Besides, in scenarios requiring immediate responses (e.g., real-time logistics or scheduling challenges), AGA can quickly generate and implement a solution, keeping up with the fast-paced demands of the situation.

Automatically generating tailored algorithms is especially beneficial in complex, dynamic, real-world situations where standard algorithms fall short. One such domain is complex logistics and supply chain management, where these algorithms can efficiently manage intricate distribution networks and optimize delivery routes under constantly changing conditions. In financial market analysis, the volatility and data-intensive nature of the markets make custom algorithms ideal for dissecting market trends, predicting stock performances, and refining investment strategies. Similarly, in the healthcare and biomedical fields, personalized medicine benefits greatly from algorithms designed to process individual patient data, offering insights for personalized treatment plans and analyzing complex genetic information.

In energy management, particularly within smart grid systems, custom algorithms are invaluable for adapting to fluctuating energy supply and demand, efficiently integrating renewable energy sources, and managing distributed energy resources. The telecommunications industry also sees potential benefits as tailored algorithms optimize network routing, manage bandwidth allocation, and improve signal processing. In the rapidly evolving e-commerce and retail sectors, custom algorithms can enhance recommendation systems, optimize inventory, and provide personalized marketing strategies based on detailed consumer behavior analysis. Furthermore, autonomous vehicles and robotics, which operate in unpredictable environments and require real-time decision-making, rely on specially designed algorithms to process sensor data and navigate safely.

Environmental monitoring and climate modeling represent another critical area where custom algorithms play a pivotal role, processing complex environmental data for accurate climate predictions, pollution tracking, and efficient resource management. The cybersecurity sector, too, relies on these algorithms for real-time threat detection, analyzing unusual patterns, and developing adaptive response strategies to evolving cyber threats. Lastly, scientific research and simulations in fields like astrophysics or quantum mechanics benefit from algorithms that can handle the simulation of complex phenomena and analyze large datasets more effectively than general-purpose tools. In each of these cases, the ability to automatically generate algorithms tailored to specific challenges leads to more effective, efficient, and accurate solutions, demonstrating this approach's significant impact across various industries and disciplines.

Developing personalized algorithms for monitoring and controlling software on personal devices is crucial in cybersecurity, particularly in protecting against unauthorized data access and transmission. Personalized algorithms for monitoring and controlling software on personal devices represent a sophisticated and user-centric approach to cybersecurity. They offer dynamic, real-time, and adaptive solutions to protect against unauthorized information access and data transmission, aligning with individual user preferences and evolving security landscapes. As cyber threats become more sophisticated, such personalized and intelligent security measures are becoming increasingly essential in safeguarding personal data and privacy in the digital age.

Constructing a model to generate an algorithm for various applications, as described earlier, is a process that essentially involves designing the necessary components to solve the master problem presented in Chapter 2, i.e., the algorithm space, the parameters space, the problem instances, the performance metrics, and the objective function.

Constructing a model to generate algorithms automatically involves a holistic approach encompassing a deep understanding of the application domain, careful design of the Master Problem components, and iterative development. This process requires a balance between theoretical understanding and practical considerations, ensuring that the developed algorithms are theoretically sound and practical in real-world scenarios. The aim is to create a system that adaptively and efficiently solves complex problems tailored to different applications' specific needs and nuances.

3.8 A FUNDAMENTAL AGA MODEL

Analogous to the AI and OR fundamental problem definitions, we can define a fundamental problem in the automatic generation of algorithms. Consider a real-world situation where the most appropriate model results in an OR, NP-hard optimization problem. No polynomial algorithm has been found for this situation, so the AGA sounds like a perfect alternative. Also, we know the mathematical formulation for the NP-hard problem as Minimize $f(x)$ subject to $x \in \Omega$ where Ω is the space of feasible solutions. In this situation, we also count some small instances with a known optimal solution for that problem.

Defining a fundamental problem in AGA for NP-hard optimization in OR is a significant step toward tackling complex real-world problems where conventional algorithmic approaches fall short. By systematically exploring the spaces of algorithms, parameters, and problem instances and by focusing on performance maximization in the Master Problem, AGA provides a structured yet flexible framework. This approach enhances the problem-solving capabilities for a specific NP-hard problem. It offers a blueprint that can be adapted to various other complex scenarios, thereby broadening the scope and impact of AGA in computational problem-solving. The following master problem components should be defined:

- **Solution container**. The solution container is a dynamic data structure that holds the current solution x during the search process. It also stores auxiliary solutions and relevant information necessary for various algorithmic operations. It forms the core upon which algorithms operate, facilitating the modification and improvement of solutions through various operations or primitive operands. There are several possibilities to implement the container in each computer language that typically consider lists or arrays depending on the characteristics of the feasible solution x.

- **Algorithm Space** (Ω_A). This encompasses a range of potential candidate algorithms for addressing the NP-hard problem, encompassing heuristic and metaheuristic algorithms, approximation methods, or even custom-developed algorithms specifically tailored to the problem. Furthermore, fundamental elements of an algorithm, such as loop cycles and end-if operators, can also be contemplated. Each algorithm within Ω_A interacts distinctively with the solution container, leveraging it to process and enhance solutions. Consequently, the architecture of the solution container needs to be versatile enough to accommodate the varied nature of the algorithms in this space.

- **Parameter Space** (Ω_{P_a}). Each algorithm in Ω_A, has its own set of parameters that influence its performance. Such parameters came from the components defined in the Ω_A.

- **Problem Instance Space** (Ω_I). This space contains a collection of small instances of the NP-hard problem for which the optimal solutions are known. These instances serve as benchmarks to evaluate the effectiveness of the algorithms generated by AGA.

- **Objective Function.** In this fundamental problem, the objective function of the Master Problem is essentially the mean squared error, calculated as the average discrepancy between the solutions produced by the algorithms and the optimal solutions for each instance, averaged across all instances. The objective function can also include a penalty function to ensure that the algorithms adhere to size and computational efficiency constraints.

Establishing a fundamental problem in AGA for NP-hard optimization marks a crucial advancement in addressing multifaceted real-world issues surpassing traditional algorithmic methodologies' capabilities. This approach's versatility lies in its systematic navigation across the realms of algorithms, parameter settings, and problem instances while anchoring its focus on optimizing performance within the Master Problem. Such a framework is not only robust and adaptable, but it also significantly augments the problem-solving potential for particular NP-hard challenges. Moreover, the generalizability of this fundamental problem model extends its utility beyond a single application, making it a versatile template for a diverse array of complex scenarios. This adaptability significantly expands AGA's applicability and influence in computational problem-solving, providing a comprehensive strategy for tackling intricate problems across various sectors, from logistics and supply chain optimization to advanced data analytics. This fundamental problem formulation in AGA paves the way for innovative, tailored solutions in fields where conventional algorithms are inadequate, reinforcing its role as a pivotal tool in modern computational research and application.

3.9 SUMMARY

This chapter explores the comprehensive process of modeling problems for the automatic generation of algorithms, focusing on NP-hard optimization problems in OR. It details the structured methodology, from identifying real-world problems to formulating them into mathematical models, emphasizing the critical role of modeling in bridging theoretical approaches to practical applications. Through exploring fundamental models in both OR and AI, the chapter underscores the versatility of AGA in adapting to diverse, complex scenarios, highlighting the synergy between mathematical modeling, algorithmic innovation, and computational implementation to solve intricate real-world challenges.

The main conclusion underscores the transformative potential of AGA in addressing real-world problems that traditional algorithms cannot effectively solve. By providing a flexible framework for exploring algorithmic, parametric, and instance spaces, AGA enables the creation of customized algorithms that significantly enhance problem-solving capabilities for specific NP-hard challenges. This approach expands AGA's applicability across various domains. It marks a significant advancement in computational problem-solving. It offers a systematic strategy for tackling complex logistics, healthcare, cybersecurity, and beyond issues, reinforcing its pivotal role in modern computational research and applications.

EXERCISES

1. Model the task of a delivery company that needs to deliver packages to several locations in a city. The goal is to find the shortest route to visit each delivery point. Formulate this practical situation using the fundamental models for the traveling salesman problem and the shortest path problem. Define the decision variables, objective function, and constraints.

2. A company must allocate a limited budget to different projects, each with a specific cost and expected profit. The goal is to maximize the total profit without exceeding the budget. Model this situation as a knapsack problem, defining the decision variables, objective function, and constraints.

3. A health center must continuously adjust its appointment schedules throughout the day as new patients arrive and treatment durations vary. Model this dynamic scenario where the goal is to continuously optimize the scheduling of patients to different doctors and treatment rooms, minimizing patient wait times and maximizing resource utilization. Define the decision variables, objective function, and constraints.

4. City Hall needs to model garbage collection management efficiently, where the amount at each collection point varies daily. The goal is to design garbage collection routes that minimize total travel distance and account for the uncertainty in the amount of garbage. Define the decision variables, objective function, and constraints.

5. Formulate an objective function for a master problem to rediscover or enhance a classical optimization algorithm like shortest path and maximum flow algorithms. Define the performance metrics and explain how the objective function will guide the search process in the automatic generation of algorithms.

6. The education department must assign students to schools based on their home locations to minimize travel distance and balance school capacities. Model this scenario using the clustering problem, where the goal is to group students into clusters (schools) such that the total travel distance is minimized and each school does not exceed its capacity. Define the decision variables, objective function, and constraints.

7. Model a company's budget allocation problem of exercise 2 for the automatic generation of algorithms. For the master problem, define the algorithm space with strategies like dynamic programming and genetic algorithms, the parameter space with variable mutation rates, and the problem instance space with varying project scenarios, aiming to minimize the mean squared error between algorithmic and optimal profits. Implement the AGA framework to explore, tune, and evaluate the resulting algorithms.

8. Model the detection of intruders in an Android system on a cell phone as a classification problem, where dynamic algorithms are automatically generated periodically to evolve and adapt to new user requirements. Define the decision variables as features extracted from the phone's usage data. The objective is to classify whether the current usage indicates an intruder or a legitimate user. Formulate the problem by defining the classification labels, the feature set, and the initial classification algorithm. Develop an objective function to maximize the classification accuracy while minimizing false positives and false negatives.

CHAPTER 4

AGA with Genetic Programming

4.1 INTRODUCTION

The idea of making computers self-programming machines comes from the time of their creation. The idea was to create calculating machines and craft entities that emulate human thought processes. The marvelous Alan Turing's work is evidence of that; he was a pioneer in this domain, conceiving the Turing test in 1950 as a measure of a machine's ability to exhibit intelligent behavior indistinguishable from that of a human. Implicit in this ambition was the idea that if a machine could think, then it should, in principle, be able to understand and generate its programming.

Self-modifying code, where a program alters its instructions while running, has been around for a while. However, true self-programming, where a computer starts from scratch and creates novel, functional programs, is a complex challenge today. Genetic programming (GP) and machine learning are steps in this direction, but they are guided by human-defined objectives and operate within frameworks designed by humans. One of the crucial milestones happened in the 1970s, thanks to the exceptionally stunning John Holland's foundational work, "Adaptation in Natural and Artificial Systems". By 1981, the field saw a practical application when Richard Forsyth evolved small tree-represented programs for crime scene classification.

Dr. John Koza operationalized and systematically implemented the field's theoretical foundations and empirical knowledge, thereby

DOI: 10.1201/9781003542896-4

formalizing and disseminating the discipline of GP. Starting with his series of books in 1992, he significantly broadened the scope and understanding of GP methodologies. His work led to the algorithmic evolution of numerous artifacts and systems, often demonstrating performance metrics that matched or exceeded those of human-engineered solutions.

GP is one of the techniques used in evolutionary computing (EC). EC is a subfield of artificial intelligence inspired by natural evolution imitating the natural laws governing biological evolution. Drawing inspiration from the model of natural evolution, it seeks to understand and replicate how living entities change and adapt over time, acknowledging the inherent transience of life, where every life form has a defined beginning and end. By emulating these principles, EC develops algorithms that evolve potential solutions over successive iterations, mirroring the continuous cycle of birth, adaptation, and termination observed in nature. This method provides a framework to solve complex optimization and search problems grounded in the very essence of life's evolutionary processes. Specifically, it employs operations like selection, mutation, recombination, and reproduction to optimize a population of potential solutions to a specific problem over multiple generations. Besides GP, EC also includes genetic algorithms and evolutionary strategies.

GP is a technique that allows for the generation of programs to solve specific tasks. It begins with a population of generally unfit programs, often randomly generated, and refines these programs using operations inspired by natural genetic processes. The core idea is to use principles of evolution, such as selection, crossover (recombination), and mutation. By continuously evaluating and selecting the fittest programs based on their performance on the desired task, GP iteratively improves the population, driving it toward solutions that are increasingly fit for the task.

Programs in GP are typically syntax tree structures, with functions or operations at the nodes and variables or constants at the leaves. Over successive generations, these trees combine among them, mimicking the evolutionary process. The method has been applied to various problems, from symbolic regression to software repair, offering the potential for innovative solutions that might elude traditional programming approaches.

The tree representation fundamental to GP offers an intuitive framework for conceptualizing and manipulating algorithms. At its core, GP treats programs as hierarchical trees, where branches represent operations or functions, and leaves correspond to input variables or constants. This inherent structure aligns closely with how algorithms manifest as nested functions and operations applied to inputs. Given this representation, GP lends itself to the domain of the automatic generation of algorithms (AGA). By evolving these trees, we essentially engage in an algorithmic evolution process. From an initial population of diverse, often random algorithmic structures, GP applies genetic operations such as crossover (tree-based recombination) and mutation (tree modifications). As the evolutionary process progresses, these trees—or algorithms—become more refined and tailored to a particular problem or dataset.

GP encompasses various forms that extend beyond the standard tree-based representation, each tailored to specific problem domains and offering unique advantages. Linear GP represents programs as linear sequences of instructions, akin to machine code, making it suitable for imperative programming tasks. Cartesian GP uses directed acyclic graphs, ideal for evolving digital circuits and graph-structured problems. Grammatical Evolution generates grammar-based programs, allowing flexibility in problems best expressed through specific languages. Stack-based GP simplifies operations with a stack-based execution model, while Graph-based GP accommodates complex relationships with graph representations. Gene expression programming combines linear chromosomes with tree structures for a versatile approach. Despite the diversity of these forms, we prefer the tree-based GP method due to its intuitive alignment with hierarchical algorithm structures.

4.2 THE MASTER PROBLEM IN AGA

Behind GP and its innumerable applications lies a fundamental optimization problem. This problem is the master problem described in Chapter 2. This intrinsic connection of GP to optimization empowers it significantly because it allows modifications to the objective function or the feasible space, enhancing GP's versatility and applicability. By accurately recognizing and modeling the underlying optimization problem, researchers can tailor GP to meet the specific requirements of diverse applications. This adaptability means GP can handle various constraints, objectives, and dynamic changes in the problem environment. For instance, by adjusting the objective function, GP can prioritize different aspects such as solution accuracy, computational efficiency, or robustness against noise.

Similarly, modifying the feasible space can help GP navigate complex solution landscapes, avoid infeasible regions, and focus on promising areas. This flexibility broadens the range of problems GP can address and allows for developing more sophisticated, effective, and efficient solutions.

By leveraging this inherent connection to optimization, GP can evolve to solve new and emerging problems, continuously improving its performance and expanding its application horizon in fields such as symbolic regression, automated programming, design optimization, and beyond. This optimization-centric approach makes GP a powerful tool for innovation, capable of generating effective solutions tailored to the specific nuances of a given problem.

The intrinsic connection between GP and optimization, coupled with the capability of representing algorithms as syntax trees, gives rise to the AGA. In this domain, GP is a powerful method for solving the master problem, which involves finding the optimal algorithm to address a specific task. By leveraging GP's ability to evolve and adapt solutions through natural selection principles, AGA systematically explores the space of potential algorithms, iteratively refining them to enhance their performance. This approach not only automates the design of algorithms but also tailors them to the unique requirements and constraints of various problems, making GP an indispensable tool in the quest for optimized, high-performing algorithms across diverse applications.

The AGA aims to discover computational strategies for specific optimization problems. The task consists of solving the master problem, for which many possibilities exist. In this chapter, we aim to approach this task using GP. Let us consider the optimization problem (4.1) in which $f(x)$, is the function to minimize knowing that x is a solution vector limited by borders expressed in Ω. For example, considering a linear integer optimization problem, the problem in (4.1) is instanced as in (4.2) where the m by n matrix A, vectors b and c are the problem instance data.

$$\text{Min}\, f(x) \text{subject to} : x \in \Omega \tag{4.1}$$

$$\text{Min } c^T x \text{ s.to}, Ax = b, x \geq 0, x \text{ integer}. \tag{4.2}$$

The master problem in Equation (4.3) considers an objective function that evaluates the performance or fitness of a candidate algorithm. In the context of algorithm generation, x represents a feasible solution for the problem (4.1). The specific algorithmic structure or logic, a denotes an algorithm and p_a, signifies specific parameter setting for the algorithm a. Finally, i is a specific set of instances occupied to produce the algorithm.

$$\Pi: \ \min F(x, a, p_a, i) \text{subject to} : x \in \Omega, a \in \Omega_A, p_a \in \Omega_{P_a}, i \in \Omega_I \tag{4.3}$$

The master problem (4.3) is a multidimensional optimization challenge encompassing multiple layers of decision variables, each with its own searching space. Tackling such a problem might be computationally challenging. Breaking it down and employing strategic simplifications, at least initially, can make the solution more tractable. The standard way to approach problem (4.3) is to choose one algorithm, for example, a metaheuristic considering a parameter setting. In this approach, we expect the produced algorithm to work well for any problem instance with the same parameter setting. Another approach is to advance with one dimension at a time. Start by optimizing over a while keeping the other two dimensions fixed. Once you've found an optimal or near-optimal solution, move to the next dimension, say p_a that is, tuning the parameters for the algorithm found in the former step considering a classification of instances. The process could repeat several times until it reaches a stop criterion.

The master problem encapsulates a multidimensional and potentially combinatorial optimization challenge. Traditional optimization techniques, which often rely on gradient information or require a convex objective, might falter when faced with the rugged landscape of such problems. This is where metaheuristic algorithms, like Genetic Algorithms (GAs), GP, Simulated Annealing (SA), and Particle Swarm Optimization (PSO), come into play. Natural processes inspire these algorithms to navigate complex search spaces without getting easily trapped in local optima.

Modeling the master problem to use such metaheuristic algorithms involves mapping the problem's variables to representations suitable for the chosen metaheuristic. For GAs, this could mean encoding these variables as chromosomes in a genome. In SA, a neighboring function would allow to perturb a current solution, while in PSO, the problem's dimensions would dictate the dimensions of a particle's position and velocity. The master problem objective function is the fitness function guiding the search process. These algorithms iteratively update potential solutions (algorithms), aiming to improve them based on the fitness function. Crucially, they incorporate mechanisms to avoid premature convergence, ensuring a broad exploration before narrowing down. While metaheuristics might not always find the global optimum, their strength lies in quickly finding "good enough" solutions, making them particularly valuable for problems as intricate as the master problem.

4.3 GENETIC PROGRAMMING

We are deepening our exploration into the field of GP, which is distinguished by its specific focus on the evolution of computer programs. At its core, the objective of GP is not just to create but to progressively refine computer programs so that they can adeptly execute specific tasks without the need for explicit, manual programming. GP employs a mechanism where a diverse population of candidate solutions, represented as programs, undergoes iterative selection, crossover, and mutation cycles. Over successive generations, this evolutionary process naturally favors those programs that exhibit better performance or fitness in solving the designated problem. As a result, the population gradually converges toward more optimized and efficient solutions. This bio-inspired methodology, mirroring the evolutionary processes found in nature, offers a dynamic and adaptive approach to problem-solving in computer science.

4.3.1 The General Algorithmic Evolutive Process

Figure 4.1 illustrates the flow and operations of a GP algorithm, capturing its iterative and evolutionary nature. The process begins with creating an initial random population, which undergoes fitness evaluation to measure how well each program solves the target problem. If the termination criterion is unsatisfied, the algorithm proceeds to the genetic operations phase. Each generation involves applying reproduction, crossover, mutation, and sometimes architecture-altering operations to generate new potential solutions. These genetic operations are applied based on fitness and specified probabilities, enhancing the diversity and quality of

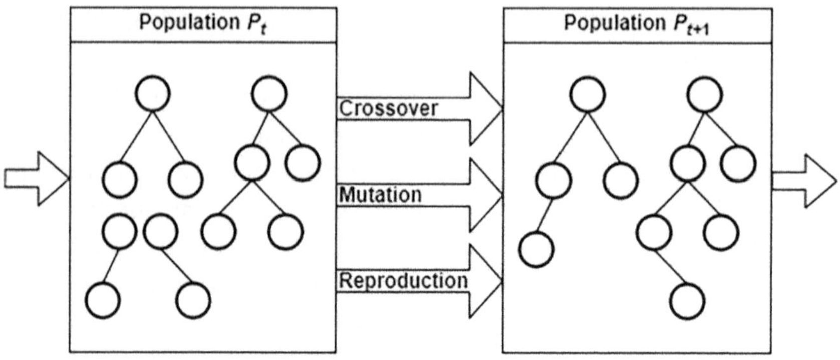

FIGURE 4.1 Schematic view of GP.

the solutions. The figure demonstrates how the generation at time t leads to the generation at time $t+1$, emphasizing the continuous refinement of programs.

In GP, the new population is formed by sequentially applying a single genetic operation—reproduction, crossover, mutation, or potentially architecture-altering operations—at each step. Reproduction involves selecting individuals based on fitness and copying them directly into the new population. Crossover selects pairs of parent individuals, exchanging segments of their code to create new offspring. Mutation introduces random changes to individuals while architecture-altering operations modify the structure of the individuals. These genetic operations are applied iteratively, with fitness evaluation guiding the selection process. The evolution continues until reaching the termination criterion, which could be a maximum number of generations, achieving a desired fitness level, or no significant improvement over several generations. This structured approach ensures efficient exploration and exploitation of the solution space, leading to optimized, efficient, and practical solutions.

GP optimizes because it seeks to evolve better solutions for a given problem over successive generations. The optimization process in GP is motivated by the pursuit of the "fittest" solutions that best meet the objectives of the problem. That is fundamentally a search process that visits the space of possible solutions to find ones that perform the best. The fitness function is fundamental to optimization because It provides a measure to rank individuals based on how well they solve the problem or meet the desired objectives. Without a fitness function, there's no way to determine which solutions are better. It sets the optimization criteria. In turn, the selection mechanism decides which individuals get to reproduce and contribute to the next generation. The selection mechanism plays a pivotal role in guiding the search toward promising regions of the solution space; its decisions are based entirely on the values provided by the fitness function. The selection mechanism doesn't operate in a vacuum; it relies on the fitness scores to make informed choices.

GP trees provide a natural and flexible way to represent hierarchical structures, making them well-suited to model expressions, decisions, loops, and other programming constructs. A node in a tree might represent an operation (like addition, subtraction, etc.), while the leaves could represent input variables or constants. Furthermore, tree structures can easily be modified, allowing for the straightforward application of genetic operations. They can grow, shrink, or change shape, enabling a vast search

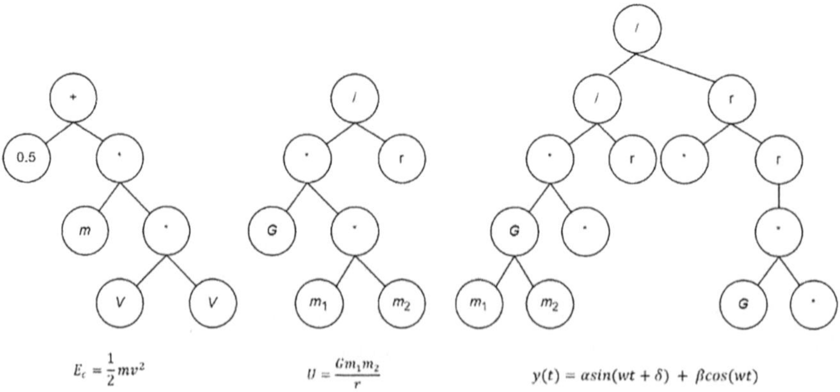

FIGURE 4.2 Tree structures representing equations.

space of potential program structures. Figure 4.2 shows three mathematical equation tree structures. The first is the kinetic energy equation in classical physics that involves the mass m and the velocity v. The second is the equation for gravitational potential energy that establishes an object's energy due to its position relative to another object under the influence of gravity, considering mass m_1 and m_2 and the gravitational constant G. The third equation represents a superposition of sine and cosine functions in time, potentially oscillating at the same angular frequency w. Combining these sine and cosine functions allows the representation of various oscillatory behaviors depending on A, B, ω, and ϕ parameters. These trees have "+" and "*" as internal nodes representing functions and constant values as leaf nodes representing terminals.

4.3.2 Genetic Operations in GP

The most common operations in a GP algorithm are selection, crossover, and mutation. By combining the tree representation with these genetic operations, GP creates, evolves, and refines a population of potential program solutions, iteratively improving their performance on the desired task.

- Selection. It's the process of choosing which programs in the current generation will be parents for the next generation. Various strategies exist, like tournament selection, roulette wheel selection, and elitism, to name a few. The fundamental idea is to probabilistically favor programs that have better fitness (i.e., perform the task better) but still give a chance to less fit programs to ensure genetic diversity.

- Crossover. This operator, also named recombination, is the primary genetic operator used in GP, which emulates biological reproduction. Two parent programs choose and swap subtrees from each other to produce two offspring. For instance, take two trees and swap a subtree from the first with a subtree from the second.
- Mutation. This operator typically involves randomly selecting a node in a tree and changing it. Eventually, this operator could replace a function, change a terminal node, or even swap out an entire subtree for another randomly generated one. Its function is to introduce genetic diversity into the population, ensuring the search doesn't get stuck in local optima.

4.4 GP AS A METAHEURISTIC

GP can be categorized as a metaheuristic since it is a strategy that offers a general-purpose problem approach to guide and facilitate the exploration and exploitation of the optimization problem solution space in diverse scenarios. The strength of GP lies in its inherent adaptability and flexibility. GP has a broad applicability spectrum, unlike some optimization techniques that are rigidly connected to specific problem structures. This characteristic makes GP versatile in solving various types of problems. Its capacity to represent solutions as tree-based structures potentially enables it to encapsulate solutions for complex problems. The true essence of GP's metaheuristic nature appears when you consider its parallels with other metaheuristic frameworks. As in any other metaheuristic approach, GP isn't about providing direct solutions but rather a framework that embeds problem-specific detail. This modular approach allows GP to maintain its generalized strategy while being adaptable to various problem domains.

When modeling a problem under the GP framework, foundational elements, similar to those in other metaheuristics, must be defined according to the problem. That includes determining the most appropriate representation of potential solutions, crafting a fitness function that accurately measures solution quality, deciding on initialization and genetic operation methods, and specifying termination criteria. These components don't just exist in isolation; they interact intricately, shaping the evolutionary dynamics and ultimately influencing the quality and efficiency of the solutions GP discovers. While GP offers a high-level blueprint, the art and science of effectively applying it lie in defining and tuning these critical elements, tailoring the general strategy to the specifics of the problem landscape. Specifically, when modeling an optimization problem using GP, the following essential elements are necessary:

- Solution representation. It is necessary to encode an optimization problem solution optimization as a tree. That involves determining the tree structures' function set (internal nodes) and terminal set (leaves or external nodes). The function set might contain mathematical operators, logical operators, or domain-specific functions, while the terminal set can have specific operations to modify the optimization problem solution. If the optimization problem has specific constraints, a method should handle and ensure that the solutions (trees) satisfy these constraints.
- Fitness Function: Define a fitness function to evaluate the performance or quality of each individual (tree) in the population. This function measures how well the solution encoded by a tree addresses the problem. Here, it is also possible to consider evaluating unfeasible solutions using an appropriate penalty function.
- Initial population: Choose a method to generate the initial population. Standard methods include the *full* method of generating individuals with a fixed maximum depth, ensuring that the generated trees are always full. Alternatively, in the *grow* method, individuals are generated with a maximum depth that is randomly determined.
- Genetic operators. In tree reproduction, there are various methods to choose trees based on their fitness. These methods comprise tournaments, roulette wheel, and rank-based selections. In addition, two trees combine to create new offspring trees, typically using subtree crossover. Mutation is another technique utilized to maintain diversity in the tree population. This operator involves introducing minor, random alterations to a tree, often substituting one subtree with a newly generated random one.
- Termination Criterion. This criterion could be a maximum number of generations, a desired fitness level, or the fitness no longer improving after a certain number of generations.
- Parameters. GP has some critical parameters to tune. These are the population size given by the number of trees in each generation, the maximum tree depth or size to avoid overly large trees, and the probabilities for crossover and mutation.

Example 4.1

The goal is to determine the optimal set of routes for a fleet of vehicles, given a set of customers with uncertain demands revealed only upon arrival. This situation involves minimizing the total expected distance traveled while ensuring all customer demands are met. Each vehicle starts and ends at the depot, and vehicle capacities must not be exceeded. Establish a model to use GP to approach this situation.

To model this situation, let us consider the following definitions:

- Nodes and Edges:
  - Let D be the depot.
  - Let $C = \{c_1, c_2, \ldots, c_n\}$ be the set of customer locations.
  - The set of all nodes is $V = \{D\} \cup C$.
  - Let E be the set of possible edges representing routes between nodes. Each edge $e_{\{ij\}} \in E$ has an associated distance $d_{\{ij\}}$.
- Stochastic Demands and Capacities:
  - Let q_i be the demand of the customer c_i, which is a random variable with a known probability distribution.
  - Let Q be the capacity of each vehicle.
- Tree Representation:
  - Represent the vehicle routes as trees $T_k = (V_k, E_k)$ for each vehicle k, with $V_k \subseteq V$ and $E_k \subseteq E$, such that T_k is a connected acyclic graph starting and ending at the depot D and visiting a subset of customers $C_k \subseteq C$.
- Objective Function:
  - Minimize the total expected distance traveled by all vehicles:

$$Minimize\, E\left[F(T)\right] = E\left[\sum_{k=1}^{m} \sum_{e_{ij} \in E_k} d_{ij}\right] \tag{4.4}$$

- Constraints:
  - Capacity Constraint: Ensure that the total demand on each route does not exceed the vehicle's capacity with high probability, considering that α is a small probability representing the risk of exceeding capacity.

$$\Pr\left(\sum_{c_i \in C_k} q_i \leq Q\right) \geq 1 - \alpha \quad \forall k, \tag{4.5}$$

- Route Validity: Each vehicle route must start and end at the depot:

$$T_k \text{ starts and ends at } D, \quad \forall k$$

- Customer Visit: Each customer must be visited exactly once. Considers that δ is an indicator function that is 1 if the customer c_i is in C_k and 0 otherwise.

$$\sum_{k=1}^{m} \delta\left(c_i \in C_k\right) = 1, \quad \forall c_i \in C \tag{4.6}$$

GP model for the problem:

- Solution Representation: Encode each vehicle route as a tree structure where:
  - The root node represents the depot.
  - Internal nodes represent stops at customer locations.
  - Leaf nodes represent the end of a route back at the depot.
- Fitness Function: Define a fitness function $E[F(T)]$ to evaluate the expected performance of each tree T in the population. Lower expected values of $E[F(T)]$ indicate better performance.

$$E\left[F(T)\right] = E\left[\sum_{k=1}^{m} \sum_{e_{ij} \in E_k} d_{ij}\right] \tag{4.7}$$

- Initial Population: Generate an initial population of random tree structures that adhere to basic route validity and expected capacity constraints.
- Genetic Operators:
  - Reproduction: Select high-fitness trees to copy into the new population.
  - Crossover: Combine subtrees from two trees to create offspring trees, ensuring routes remain valid.
  - Mutation: Introduce random changes to trees, such as swapping customer visits or adding/removing routes, ensuring constraints are still met.
- Termination Criterion: The evolution process terminates when a set number of generations is reached, a pre-specified fitness threshold is achieved, or when improvements between generations fall below a specific rate.
- Parameters: Tune critical parameters, including population size, maximum tree depth, and probabilities for crossover and mutation.

Example 4.2

Given an electric distribution system with a main substation and multiple consumer nodes, the goal is to design a distribution network that efficiently delivers electricity from the primary substation to all consumer nodes. The design should minimize the total electrical losses, installation, and maintenance costs and ensure the reliability and robustness of the network.

To model this situation, let us consider the following definitions:

- Nodes and Edges:
  - Let $N = \{n_1, n_2, \ldots, n_k\}$ be the set of consumer nodes.
  - Let $S = \{s_1, s_2, \ldots, s_m\}$ be the set of substations with s_0 being the primary substation.
  - The set of all nodes is $V = S \cup N$.
- Connections:
  - Let E be the set of possible edges representing connections between nodes. Each edge $e_{ij} \in E$ has an associated installation cost c_{ij} and electrical loss l_{ij}.
- Tree Representation:
  - Represent the network as a tree $T = (V_T, E_T)$ with $V_T \subseteq V$ and $E_T \subseteq E$ such that T is a connected acyclic graph where V_T includes the main substation s_0 and all consumer nodes in N.
- Objective Function:
  - Minimize the total cost $C(T)$ and electrical loss $L(T)$ of the network:

$$\text{Minimize } F(T) = \alpha \sum_{e_{ij} \in E_T} c_{ij} + \beta \sum_{e_{ij} \in E_T} l_{ij}, \tag{4.8}$$

  - where α and β are weighting factors that balance the importance of costs and losses.
- Constraints:
  - Connectivity: Ensure that each consumer node $n_i \in N$ is connected to the primary substation s_0 either directly or through other substations $s_j \in S$.
  - Reliability: Incorporate redundancy where possible to maintain supply during failures.
  - Capacity: Ensure that substations and connections do not exceed their capacity limits.

GP model for the problem:

- Solution Representation: Encode the distribution network as a tree structure where:
  - Internal nodes represent substations (including the main substation).
  - Leaf nodes represent consumer nodes.
  - Edges represent electrical connections between substations and consumer nodes.
- Fitness Function: Define a fitness function $F(T)$ that evaluates the performance of each tree T in the population considering that lower values of $F(T)$ indicate better performance.

$$F(T)=\alpha\sum_{e_{ij}\in E_T} c_{ij}+\beta\sum_{e_{ij}\in E_T} l_{ij} \tag{4.9}$$

- Initial Population: Generate an initial population of random tree structures that adhere to basic connectivity requirements.
- Genetic Operators:
  - Reproduction: Select high-fitness trees to copy into the new population.
  - Crossover: Combine subtrees from two trees to create offspring trees.
  - Mutation: Introduce random tree changes, such as adding or removing connections.
- Termination Criterion: The evolution process terminates when a set number of generations is reached, a pre-specified fitness threshold is achieved, or improvements between generations fall below a specific rate.
- Parameters: Tune critical parameters, including population size, maximum tree depth, and probabilities for crossover and mutation.

These examples show that it is necessary to represent a solution to the optimization problem as a tree to use GP as a metaheuristic. This tree-based representation is natural and intuitive in many situations, as demonstrated by the examples of electrical distribution networks and stochastic vehicle routing problems. In these cases, tree representations' hierarchical and structured nature aligns well with the problem requirements. However, for other optimization problems, this condition might not be as straightforward. More elaborate representations may be needed in such scenarios to adapt the problem to a tree-based structure effectively. This adaptation can involve transforming the problem into a hierarchical format or decomposing complex solutions into simpler sub-components that fit a tree-like structure.

4.5 MODELING THE MASTER PROBLEM WITH GP

The master problem aims to determine the optimal algorithm for the optimization problem defined in Equation (4.1). When GP approaches the master problem, it lends itself to a natural representation through tree structures. At its core, the essence of any optimization problem algorithm occurs as a series of decisions, operations, and evaluations, each of which can be represented as nodes and branches of a tree. This tree-based representation becomes particularly advantageous. Trees, as data structures, can encapsulate hierarchical relationships and processes, which align seamlessly with the inherent flow of algorithms. In the case of the master problem, the primary challenge is to determine an optimal sequence of operations. When broken down, each decision point, operation, or function becomes a node in the tree, branching out to subsequent operations. Specifically for the master problem, it is necessary to define the representation of potential algorithms, design a fitness function that measures optimization solution quality, decide on initialization and genetic operation methods, and specify termination criteria.

4.5.1 Solution Representation as a Tree

The solution representation for the master problem considers the codification of an algorithm as a tree. We can identify two sorts of nodes. The internal nodes are functions typically in an algorithm, like loops, logical operators, and conditionals. Such nodes serve as the building blocks of algorithms guiding the flow of operations. The second group of nodes corresponds to terminals or leaves, representing the final, actionable items. For the master problem, these terminals can be heuristics, which provide approximate solutions, modification operations that refine or adjust existing solutions, or even exact methods that guarantee optimality. By having such a wide array of terminals at its disposal, GP can explore a rich solution space, combining, for example, the speed of heuristics with the precision of exact methods.

a) **The tree's internal nodes**

The internal nodes, known as functions, play a crucial role in determining the structure and behavior of the evolved algorithm. When we think of an algorithm, we can imagine a set of step-by-step instructions executed in a specific order to solve a problem. The flow of these instructions is often not linear; there are decisions to be made, repetitions to perform, and conditions to check. Here is where elements like loops, logical operators, and conditionals come into play.

- Loops. Loops in an algorithm enable repeated execution of specific steps. They are helpful for cycling through data structures, progressively improving a solution, or conducting an operation until it reaches a particular condition. An internal node in a loop determines the execution frequency of its subordinate nodes, which can also be loops themselves. These iterations might occur a set number of times or continue until they fulfill a specific condition. It's essential to define the parameters related to these conditions at the start of the process.
- Logical Operators. Logical operators like AND, OR, and NOT are used to make decisions based on multiple criteria. They can combine different conditions and produce a single Boolean outcome, which can guide subsequent operations. Nodes representing logical operators can combine the outcomes of their child nodes to decide the flow of the algorithm. For example, an AND operator could have two child nodes representing conditions, and the operations following this node would only occur if both conditions are True.
- Conditionals. Conditionals such as IF-THEN or IF-THEN-ELSE allow the algorithm to branch and make decisions. A conditional executes different sets of operations based on whether a condition is True or False. An internal node representing a conditional would have child nodes representing the condition to check and the operations to execute for both outcomes (true and false). The flow of the algorithm would branch based on the outcome of the condition.

The power of the tree's internal nodes lies in their ability to dictate the flow of operations in the evolved algorithm. The tree's structure (defined by the arrangement of internal and leaf nodes) will determine the algorithm's sequence, repetition, and decision-making process. For instance, a tree might start with a loop node with a condition node as a child. This structure would mean that the algorithm first checks the condition and, based on the outcome, repeatedly executes a set of operations. Deeper nodes in the tree would further define the specifics of these operations.

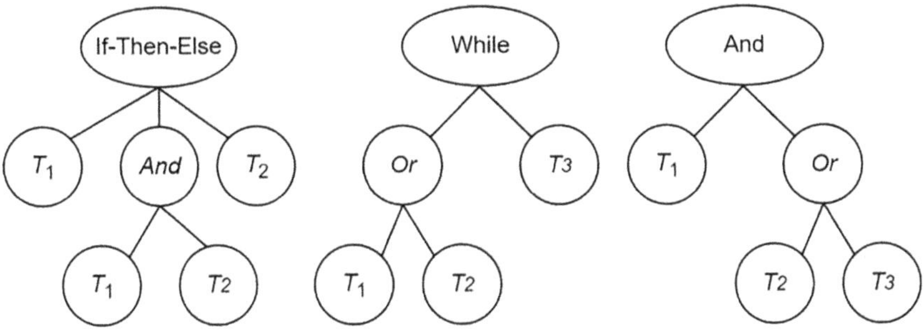

FIGURE 4.3 Three algorithmic trees.

Furthermore, as the GP evolves the tree structures over generations, the placement and type of these internal nodes can change, leading to different algorithmic behaviors and allowing the GP to explore various algorithmic strategies to seek the most effective solution to the problem at hand. Thus, the tree's internal nodes are the architects of the algorithm. They shape the structure, determine the flow, and enable the complex decision-making and iterative processes that algorithms often require.

Figure 4.3 presents three small trees representing algorithms using the set of terminals T_1, T_2, T_3 and a set of functions If – Then – Else, While, And, Or. In the first tree (a), the node If – Then – Else checks a condition. If T_1 is evaluated as True, the *And* operation is executed with T_1 and T_2. If T_1 is evaluated as False, then T_2 is executed. The second tree (b) in the figure considers a While node continuously executing if the condition within the Or operation is evaluated as True. The Or operation checks if either T_1 or T_2, is true. If the Or condition evaluates as True, then T_3 is executed, and the loop continues. If it evaluates as false, the loop terminates. In the third tree (c), the And node checks if both its children nodes are True. The left child is simply T_1. The right child is an Or operation that checks if either T_2 or T_3 is True. For the entire tree to be evaluated as True, T_1 must be true, and either T_2 or T_3 (or both) must be true. These trees provide a basic visualization of how algorithmic structures can be formed using a combination of decision-making internal nodes and actionable leaf nodes. The trees can be parsed and executed to make decisions based on the conditions set by the internal nodes and the actions associated with the leaf nodes.

b) **The tree leaf nodes**

The GP tree representation for the master problem can include a diverse set of terminal nodes. Such nodes offer a powerful, flexible approach to solving the problem in Equation (4.1). Each time the terminal works, it receives an initial solution corresponding to the algorithm's current solution stored in a solution container. When it finishes, it returns a new solution that eventually differs from the current one. Equivalently, it modifies the solutions container. Terminal nodes can be considered heuristics, metaheuristics, modification operations, and exact methods with time constraints. The resulting algorithms are not just solutions but also provide valuable insights into the problem's nature and potential strategies for tackling similar problems in the future.

Unlike the internal nodes, which generally denote operations, functions, or logical controls that determine the flow or sequence of events, leaf nodes carry specific, often tangible, values or actions on the current solution of the problem in Equation (4.1). When we run a tree, the terminal nodes provide the concrete data or instructions that feed into the higher-order operations denoted by the internal nodes. While internal nodes dictate an order in which different tree components should interact, leaf nodes determine the actions to modify the current solution.

The solution container plays a crucial role in the optimization process, serving as a dynamic repository for the evolving solution. As the algorithm progresses through its iterations, the solution container continuously updates to reflect the most recent state of the optimization problem. This container acts as a central point of reference, allowing various algorithm components, including the terminal nodes in the GP tree, to access and modify the current solution. By maintaining this mutable structure, the solution container facilitates the exploration of the solution space, enabling the algorithm to improve upon previous results incrementally. When the optimization process concludes, the solution container holds the final output. This solution represents the algorithm's best attempt at solving the problem, whether near-optimal or optimal. This mechanism streamlines the optimization process and clearly records the solution's evolution throughout the algorithm's execution.

The leaf nodes can make tangible changes to the solution container by design. Each terminal node is essentially a predefined operation or instruction set that, when invoked, knows how to modify the solution container in a specific way. For example, a terminal node might represent an instruction to swap two elements in the solution container, add an element, remove an element, or perform more complex operations like running a local search procedure. As the algorithm progresses, these terminal nodes, under the guidance of the internal nodes, collectively modify the solution container's contents. The dynamic and often synergistic interplay between terminal nodes allows diverse solution space exploration. Over time, and through many iterations, this leads to the evolution of better and more efficient algorithms. In this framework, the solution container isn't just a passive holder but a central stage where the actions of the terminal nodes come to life, resulting in the algorithm's intended outcomes.

Figure 4.4 presents a conceptual visualization of a solution container in the context of the AGA for the problem in Equation (4.1). In this instance, the solution container holds the current solution, represented as x_c, which could be a sequence or array of elements. Operations or actions performed by terminals modify or improve such solutions. The figure depicts the current solution as a linear sequence of blocks, each potentially representing variables, attributes, or components of the optimization problem in Equation (4.1). Two primary operations, "swap" and "shift" are illustrated: Swap (swap(x_c)) interchanges the positions of two components or elements within the current solution. In turn, shift(x_c) modifies the position of a single element, moving it either left or right within the solution. This operation can adjust and refine the solution's configuration, especially when order matters. The if – then operation, depicted above the specific swap and shift functions, suggests a decision-making structure governing when and how these operations are applied. This conditional approach ensures that operations are executed based on specific criteria or conditions, making the modifications more strategic and context-aware. The small tree structure embodies the operational flow of the algorithm on the current solution.

Figure 4.4 encapsulates the dynamic interplay between a solution and the operations in the optimization process. By leveraging a tree structure with decision-making capabilities and specific operations, the algorithm can systematically modify and refine the current solution, navigating

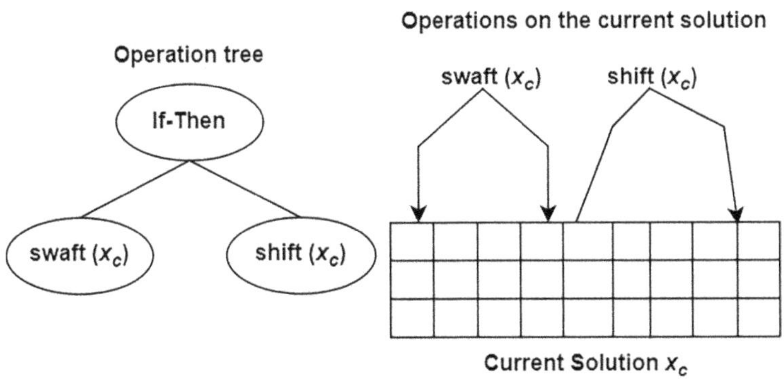

FIGURE 4.4 A representation of a solution container.

toward an optimal or near-optimal solution. This visualization reinforces the notion that algorithmic optimization is not just about finding a solution but iteratively improving upon it through strategic operations.

c) **Parsing the tree**

Given a tree representing an algorithm, it is crucial to process it correctly to execute its instructions. This systematic traversal and interpretation of the tree's nodes and their relationships is precisely where the parsing process comes into play. Parsing ensures that each instruction or operation represented by a node in the tree is executed in the correct sequence and context, allowing the algorithm or instruction set the tree represents to function as intended. In general, parsing, in computer science and linguistics, refers to the process of analyzing a sequence of symbols, often in the form of text, to determine its grammatical structure concerning a given formal grammar. This process helps understand the input's exact meaning and can transform it into a format more suitable for further processing. In programming and computing contexts, parsers interpret code, data files, or command inputs, breaking them down into a structure that software can understand and work with.

One typically performs a depth-first traversal (or a similar traversal) to parse a tree representing an algorithm. Since the tree nodes contain operations or functions (for internal nodes) and terminals or actionable items (for leaf nodes), parsing the tree means executing those operations in the order they appear during the traversal. Algorithm 4.1 uses a depth-first traversal to parse a given tree.

4.5.2 The Fitness Function

Defining an objective function to quantify the algorithmic performance of the master problem is critical. In algorithm design, assessing the solution quality or fitness relative to the optimal solution is imperative. This quantitative evaluation guides the search trajectory and evolutionary optimization process toward superior solutions. The primary objective is to minimize the deviation between the algorithm-generated solution and the true global optimum. This optimization criterion is a heuristic measure for solution quality, facilitating the comparison and ranking of candidate algorithms within the solution space. The master problem objective function acts as a fitness landscape, where the topology informs the optimization process's search direction and convergence rate. By rigorously formulating this function, we establish a mathematical framework for systematically exploring the algorithm design space and iteratively refining the solution.

Algorithm 4.1: Parse_Tree(node)

```
Input: node; current solution;
Output: Current solution updated
Begin
   If node is NULL: Return
   If node is a leaf node (terminal):
      Execute the operation represented by the node;
        Return;
   Execute the node function before exploring its
     children;
   For each child of node: Parse_Tree(child);
End.
```

Researchers must test an algorithm on various problem examples to assess how well it performs on different optimization problems. This test set should include many problem instances to ensure a thorough evaluation. However, precise quantification of algorithmic performance necessitates a priori knowledge of the global optima for these instances. With these benchmark optima established, we can rigorously compute algorithm-generated solutions' deviation or error metrics. Consequently, utilizing a curated dataset of problem instances with known optimal solutions makes it feasible for an algorithm to demonstrate near-optimal convergence on these test cases. This performance data is a fitness criterion that informs GP operators to explore the algorithmic search space. The resultant fitness landscape guides the meta-optimization process, facilitating the identification of high-performing algorithmic structures and parameterizations.

The optimization landscape is vast and riddled with nuances. When devising an algorithm to tackle an optimization problem, it's paramount to ensure that it isn't narrowly tailored to a specific subset of instances but is robust and generalizable across a diverse range of scenarios. Having a set of problem instances with known optimal solutions serves as a benchmark. These benchmarks enable us to gauge an algorithm's proficiency quantitatively. By comparing the algorithm's output with these known optima, we can discern the degree of deviation, thereby obtaining a tangible measure of the algorithm's efficacy. Such benchmarks provide a grounded, empirical foundation, ensuring that our algorithm doesn't merely function in theory but thrives in practice.

Furthermore, GP leverages these benchmarks through its evolutionary operators. When testing an algorithm against various problem instances and finding the resultant errors, GP uses this feedback to guide its search for better algorithms. For instance, algorithms that yield solutions closer to the optimal might be prioritized and propagated to subsequent generations through crossover and mutation operations. Conversely, those that deviate substantially might face "extinction". By continuously challenging and evaluating potential solutions against known benchmarks, GP iteratively refines the algorithmic candidates into ones that exhibit precision and versatility. The evolutionary paradigm of GP, thus, thrives on feedback, using the known optima of benchmark instances as guideposts to navigate the search space effectively.

An error estimation measures the performance of an algorithm. Two standard metrics are the Mean Square Error (MSE) and the Mean Absolute Value Error (MAPE). Let s_i^*, be the optimal value of the problem instance i and s_i, the value found by an algorithm π. We also have m problem instances with a known optimal value. Let be f_π the objective function of the master problem.

- MSE. The MSE computes the average squared difference between the estimated and actual values for a given set of problem instances. MSE amplifies the impact of more significant errors over smaller ones due to the squaring operation. Thus, it's more sensitive to outliers or significant deviations from the optimal solution. Using MSE as the objective function would prioritize algorithms that reduce these more significant errors. Equation (4.10) presents the MSE. This equation accumulates the squared differences between the optimal values and the values found by the algorithm for each instance. It then takes the average by dividing it by the total number of instances m. The objective of the master problem would be to minimize this f_π value, which represents the average squared error across all problem instances.

$$f_\pi^{\text{MSE}} = \frac{1}{m}\sum_{i=1}^{m}\left(s_i - s_i^*\right)^2 \tag{4.10}$$

- MAPE is a measure used to determine the accuracy of a forecasting method in predicting values. It expresses the error concerning the actual values. Equation (4.11) computes the absolute percentage error for each instance, sums them up, and then takes the average by dividing by the total number of instances. The objective would be to minimize the f_π^{MAPE} value to get the algorithm that produces solutions closest, on average, to the optimal solutions in terms of percentage error. Unlike MSE, it treats all errors the same, whether large or small. That can be beneficial when minimizing the average error without emphasizing occasional significant errors.

$$f_{\pi}^{\text{MAPE}} = \frac{100}{m} \sum_{i=1}^{m} \left| \frac{s_i - s_i^*}{s_i^*} \right| \tag{4.11}$$

When deciding between these two metrics, it's essential to consider the nature of the master problem and the implications of the type of error. When larger deviations from optimality are much less acceptable than smaller ones, using MSE might be more suitable. On the other hand, if all errors, regardless of size, are equally problematic, the MAPE would be a better choice. The chosen error metric will guide the evolutionary process. Algorithms that produce solutions with lower error values (according to the chosen metric) will be deemed more "fit" and more likely to be selected and reproduced in subsequent generations, guiding the population toward increasingly effective algorithmic solutions.

In the context of GP and optimization, penalty functions are crucial in guiding the search for optimal solutions by incorporating constraints into the objective function. The objective function of the master problem can include convenient penalty functions to manage constraints more effectively and ensure the algorithm adheres to specific requirements. Penalties can address various issues, such as the size of the algorithms, the frequency with which the algorithm fails to find the optimal solution, or the undesirable phenomenon of bloating, where the program grows excessively large without corresponding performance improvements. By integrating these penalty functions into the objective function, the optimization process can penalize solutions that violate constraints or exhibit undesirable traits. This approach helps balance exploration and exploitation, ensuring that the evolutionary process remains focused on generating efficient, effective, and feasible solutions. Penalty functions thus enhance the flexibility and applicability of GP, allowing it to handle a broader range of complex optimization problems by effectively managing constraints within the optimization framework.

Penalty functions must be weighted appropriately to ensure that the optimization problem remains focused on finding a good algorithm for the optimization problem at hand. These penalty terms in the objective function introduce new adjusting parameters, which require careful tuning to balance the impact of penalties with the primary objective. A general guideline is to ensure that the main objective, such as the algorithm's performance, holds over 90% of the weight in the objective function. This threshold ensures that the search process prioritizes finding effective solutions while considering penalties. By appropriately weighting the penalty functions, the evolutionary process can efficiently navigate the solution space, ensuring the fulfillment of the constraints without detracting from the primary goal of optimizing algorithm performance. This balanced approach helps maintain the integrity and effectiveness of the GP methodology in solving complex optimization problems.

4.6 THE EVOLUTION OF ALGORITHMS

Having defined a solution representation and the fitness functions to guide the evolutionary process, it is now possible to solve the master problem and produce the best algorithm for the optimization problem in Equation (4.1). Figure 4.5 illustrates the evolutionary process of algorithms, embodying a central principle behind evolutionary computation. At its core, this process draws inspiration from natural evolution, applying reproduction, crossover, and mutation to evolve populations of algorithms capable of finding optimal or near-optimal solutions. Each tree within the "cloud" or population signifies an individual algorithm, depicted in tree structure form. The reproduction operator allows algorithms from the current population to pass directly to the next generation, typically based on their fitness. The more fit an algorithm is (i.e., how well it performs its intended task), the higher its chances of being selected for reproduction. In

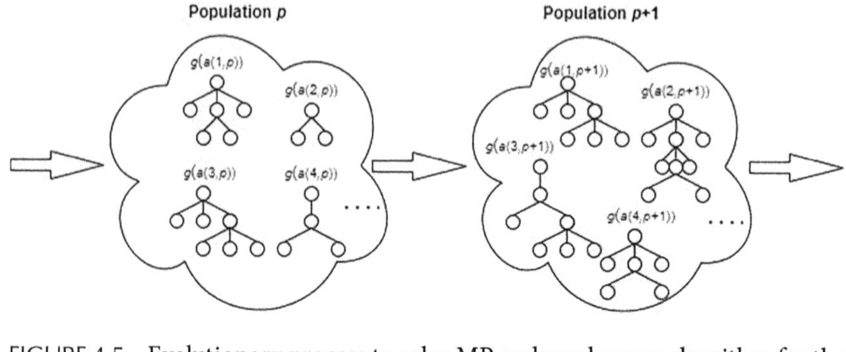

FIGURE 4.5 Evolutionary process to solve MP and produce an algorithm for the problem (4.1).

addition, the crossover operates with two algorithms (parents) exchanging parts of their structures to produce one or more offspring. That can help combine the two algorithms' best parts to produce a more effective offspring algorithm. Finally, the mutation introduces small, random algorithm structure changes. It ensures diversity in the population, preventing premature convergence to a suboptimal solution and allowing exploration of new areas in the solution space.

After applying the genetic operators, a new population of algorithms arises. This new generation might include a mix of reproduced algorithms, offspring from crossover, and mutated versions of the original algorithms. Ideally, as the evolutionary process progresses through multiple such cycles, the population's overall fitness should increase, leading to the emergence of increasingly practical algorithms. Thus, the arrows between the populations highlight the continuous nature of this evolutionary process. The new population serves as the starting point for the next iteration of evolution. Over multiple iterations, the evolutionary process refines the algorithms in the population, seeking to optimize their performance for the task at hand. Thus, Figure 4.3 captures the dynamism and adaptability of the evolutionary process in algorithm development. By mimicking the principles of natural evolution, it offers a robust method to discover high-performing algorithms, even in complex and ever-changing problem domains.

Example 4.3

Let us consider the binary knapsack problem (BKP). It is a classic optimization challenge. The objective is to determine a subset of items that can be accommodated within a knapsack while maximizing the total profit derived from those items. This problem can be precisely formulated as an integer programming problem, adhering to Equations (4.12)–(4.14). The binary decision variable x_j, representing whether an item j is selected ($x_j = 1$) or not ($x_j = 0$) for inclusion in the knapsack. Thus, the objective is to determine the optimal set of x_j that maximize the total profit while respecting the knapsack's capacity constraint. Here, W represents the positive capacity constraint of the knapsack, while n is the total number of available items. Each item under consideration possesses two vital attributes: a positive profit value denoted as p_j and a corresponding positive weight value denoted as w_j, where j ranges from 1 to n. Furthermore, it is assumed that p_j, w_j, and W are integer values, emphasizing the discrete nature of this problem-solving scenario. We look for a model of the master problem to find algorithms able to find near-optimal solutions for BKP, considering the set of functions $F = \{\text{If} - \text{Then}, \text{While}, \text{Or}, \text{And}\}$.

$$\text{Maximize } Z = \sum_{j=1}^{n} p_j x_j \tag{4.12}$$

$$\text{Subject to :}$$

$$\sum_{j=1}^{n} w_j x_j \leq W \tag{4.13}$$

$$x_j \in (0,1), j = 1,\ldots,n \tag{4.14}$$

To the modeling of the master problem to automatically generate algorithms for BKP using GP, we can define the building blocks:

a) Solution Container for KP: A solution container SC for the KP is a binary string of length n corresponding to the number of items. Each position j in the string indicates whether item j is included (1) or excluded (0) from the knapsack.
b) Functions Set F
    - If-Then: Conditional operation to check a condition and perform an action. Specifically, If-Then (P_1, P_2) executes operand P_2 only if operand P_1 returns True. The function returns True if P_1 returns True.
    - Or: Logical operation to evaluate if at least one of two conditions is true. Specifically, Or (P_1, P_2) executes operand P_2 only if P_1 returns False and returns True when P_1 or P_2 has returned True
    - And: A logical operation to evaluate if both conditions are True. Specifically, And (P_1, P_2) executes both operands and returns True if both operands also return True. Otherwise, it returns False.

- While: Loop operation repeatedly working until a condition is met. Specifically, While (P_1, P_2) executes the terminal or function P_2 as long as P_1 continues to return True. Then, the process repeats until P_1 returns False. In addition, an upper limit to the running time or number of repetitions is necessary to avoid infinite operation.

c) Terminals: Given a current BKP solution in the solution container (SC), the total weight used SC.*weight* and the BKP objective value SC.*f* can be simultaneously stored. Thus, we define four terminals.

- T_1 (Add-Item): Add the next available item to the knapsack if its weight doesn't exceed the capacity. The pseudo-code is:

```
T1 (SC, Flag): Add a new item.
  Input: Current SC, SC.weight, SC.f;
  Output: Updated SC, SC.weight, SC.f;
  Current_f = SC.f ; Flag = False;

Begin
  If there are unselected items Then
          Pick the next unselected item i;
          If weight of i + SC.weight ≤ W Then
          Add item i to the SC;
          Update SC.weight, and SC.f;
          If SC.f≥ Current_f Then Flag = True;
End.
```

- T_2 (Remove-Item): Remove item from the knapsack. The pseudo-code is:

```
T2 (SC, Flag): Remove an item.
Input: Current SC, SC.weight, SC.f;
Output: Updated SC, SC.weight, SC.f;
Current_f = SC.f; Flag = False;
Begin
  If there are selected items Then
          Pick the next selected item i
          Remove item i from the SC
          Update SC.weight and SC
          If SC.f ≥ Current_f Then Flag = True
End.
```

- T_3 (Add-Best-Value): Add to the knapsack the best-valued item if the new item's weight doesn't exceed the remaining capacity. The pseudo-code is:

```
T3 (SC): Add best valued item.
Input: Current SC, SC.weight, SC.f;
Output: Updated SC, SC.weight, SC.f;
Current_f = SC.f; Flag = False;
Begin
  If there are unselected items Then
            Pick an unselected item i with the
                highest profit-per-weight
                ratio;
            If weight of i + SC.weight ≤ W Then
               Add item i to the SC;
               Update SC.weight and SC;
       If SC.f ≥ Current_f Then Flag = True
End.
```

- T_4 (Remove-Worst-Valued): This terminal removes the worst-valued item according to the profit-weight ratio.

```
T4 (SC): Remove the item with the worst value.
Input: Current SC, SC.weight, SC.f;
Output: Updated SC, SC.weight, SC.f;
Current_f = SC.f ; Flag = False;
Begin
     If there are selected items in SC Then
     Pick a selected item i with the lowest
         profit-per-weight ratio;
     Remove item i from the SC;
     Update SC.weight and SC;
End.
```

d) Objective Functions for the Master Problem

For the objective function, Equation 4.10 ensures the minimization of the error incurred by an algorithm when it evaluates a given set of BKP instances. Thus, to apply AGA to BKP, it is necessary to dispose of a set of instances with a known optimal value.

Having defined the building blocks to generate algorithms for the binary knapsack problem, we can identify some possible algorithms resulting from such basic definitions. Figure 4.6 shows two possible algorithms. The two tree structures represent different strategies for the

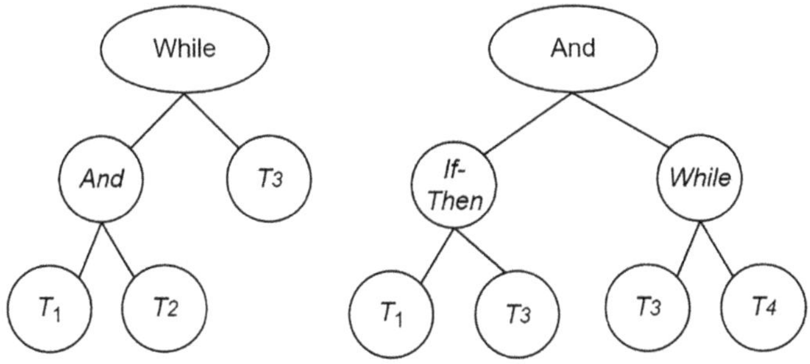

FIGURE 4.6 Two possible algorithms to find near-optimal KP solutions.

knapsack problem. In the first tree, the algorithm consistently adds two items to the knapsack, provided they don't exceed the weight constraint, iterating this process and removing one item in each iteration. The second tree first tries to include a new item by terminals T1 or T3. Concurrently, it prioritizes adding the best-valued items while continuously swapping out the least valuable item in the knapsack, ensuring its content is optimized. Both strategies provide a unique approach to maximizing the value within the knapsack's weight constraints.

4.7 CONSTRUCTIVE AND REFINEMENT ALGORITHMS

There are many ways to design functions, terminals, and containers. Depending on the initial choice, we can obtain refined, constructive, or hybrid algorithms. Let us consider the two extreme ways. If the solution container is initially empty and the set of terminals contains actions allowing the increases of the current solution, then the evolutionary process will return constructive algorithms. This strategy means that the algorithms can construct BKP solutions from scratch. In addition, if the initial solution is feasible and the remaining capacity does not admit a new item in the knapsack, and we also consider terminals that modify the solution, new refinement algorithms may appear at the end of the evolutionary process. When both types of terminals—constructive and single-solution—are present, we can obtain hybrid algorithms. The production of such hybrid algorithms combining constructive and single-solution approaches is feasible under these conditions.

This analysis encapsulates the potential dynamism of evolutionary algorithms in addressing optimization problems. When starting from an empty solution container, the constructive approach steadily builds toward an optimal or near-optimal solution using the provided terminals. This incremental approach mirrors how many heuristics operate, systematically adding elements based on specific criteria until they reach the end condition. On the contrary, beginning with a feasible solution and iterating on it by potentially improving or refining it with each step aligns with the concept of local search or refinement heuristics. Combining constructive and refinement modifying strategies might introduce various solution paths. It has the potential to harness the best of both worlds: the wide exploration breadth of constructive algorithms and the deep exploration depth of single-solution refining ones. The hybrid approach would offer flexibility and adaptability, catering to different problem instances or diverse constraints. It's reminiscent of how nature itself often finds solutions, not just through gradual construction but also through constant refinement and adaptation. The balance between exploration (finding new solutions) and exploitation (refining existing solutions) becomes pivotal in such a hybrid strategy.

The constructive and refinement algorithms that would emerge with the appropriate terminal and function sets could work on constructing and reconstructing the optimization problem solution. In such a way, looking for the optimal solutions for a given instance works with ups and downs. The master problem objective function oversees the selection of only algorithms that effectively solve the optimization problem. Thus, a final algorithm works efficiently with ups and downs. This process resembles human behavior in finding a suitable solution for daily situations.

The metaphor of the ups and downs in the algorithmic process echoing human problem-solving behavior is spot on. In real life, we often don't follow a linear path to find a solution; instead, we iterate, backtrack, and refine our strategies based on the feedback we receive. Similarly, combining constructive and refinement algorithms offers a dynamic and adaptive approach to combinatorial optimization. Allowing the algorithm to construct and then deconstruct (or reconstruct) solutions captures the essence of iterative problem-solving. That offers a more extensive exploration of the solution landscape and fosters adaptability. The master problem's objective function is a guiding hand, akin to our cognitive processes, ensuring the chosen strategies are effective. This holistic approach might yield more robust and efficient solutions reminiscent of humans' heuristic strategies in their day-to-day decision-making.

4.8 ROBUSTNESS VERSUS SPECIALIZATION

If we select a variety of problem instances to generate algorithms for combinatorial optimization problems, we can expect the generation of robust algorithms; algorithms can find near-optimal solutions for practically any problem instance. This scheme follows the paradigm in science and engineering: looking for only one algorithm to solve any problem instance. The underlying principle here is like the search strategy in machine learning, where diverse training data leads to better generalization in models.

The challenges of combinatorial instances require specific metrics to capture their inherent complexity and variety accurately. Standard metrics include instance size, which directly affects the exponentially growing search space; constraint tightness, which measures the ratio of constraints to variables; and instance density, which evaluates the proportion of potential connections. Objective function landscapes can offer insights into the problem's ruggedness with peaks, valleys, and plateaus, while the distance to the optimal solution gauges the problem's difficulty. The gap between the best and worst solutions can hint at the diversity of the solution space. Depending on the specific problem, unique metrics might emerge, like the distribution of city distances for the Traveling Salesman Problem. Utilizing a combination of these metrics allows for a comprehensive understanding of instance variety, paving the way for the generation of robust algorithms that can tackle diverse challenges in combinatorial optimization.

The other extreme of robustness is the specificity of algorithms for each instance. In that case, each instance needs a particular algorithm. This fact gives rise to a very different paradigm from what is usual in algorithm design. If each instance in the space of instances needs a specific and different algorithm, this could explain the difficulty in consistently finding optimal solutions across various instances using a single algorithm.

Traditionally, the goal has been to design universally applicable algorithms to tackle many problem instances. However, tailoring an algorithm for each specific instance introduces a new dimension of granularity in problem-solving. If every problem instance has unique quirks and nuances, a bespoke algorithm could exploit these peculiarities to find optimal solutions more efficiently than a one-size-fits-all method. The traditional approach seeks a balance between generality and performance, potentially sacrificing the latter. Tailoring algorithms for each instance may be resource-intensive but could ensure consistently high performance. It shifts the emphasis from designing universally adaptable algorithms to creating an adaptive system capable of crafting bespoke algorithms for each unique problem instance. This concept challenges established norms and offers a fresh avenue for exploration in the realm of heuristic and metaheuristic research.

In this approach, an algorithm is tailored specifically for a single instance, achieving optimal or near-optimal solutions for that particular case. It becomes redundant once it has fulfilled its purpose because it is over-specialized in solving the problem instance. This concept underscores a shift from seeking universally applicable solutions to creating hyper-specialized tools that serve a singular purpose efficiently but become obsolete afterward. This concept challenges traditional computational wisdom and opens avenues for exploring the efficiency and viability of such algorithms. Such an idea generates the notion of a "disposable" algorithm. In many ways, this is analogous to the function of a capsule coffee machine. Each capsule, precisely measured and designed for a single use, contains the ideal blend and quantity of coffee to brew just one perfect cup. The capsule is not valid anymore once the coffee is ready, fulfilling its unique purpose. It's not reusable for another round or adaptable for a different kind of beverage. Analogously, the disposable algorithm has a one-time, specific use. Much like you wouldn't attempt to repurpose a spent coffee capsule, you wouldn't redeploy a disposable algorithm for another instance. The coffee capsule and the disposable algorithm highlight a paradigm where efficiency and precision for a singular task take precedence over reuse and general applicability.

Let us revise the intermediate case. We look for overfitting algorithms for a class of instances. We can initially separate the set of instances into different classes. For example, we can consider dense or sparse graphs in the Traveling Salesman Problem. We can even cluster optimization problem instances based on instance features. For example, we can use a clustering algorithm to prepare k groups of instances. Then, we overfit the AGA for each group. Thus, overfitting in AGA is not a problem. Now, it become the specialization of algorithms. So, we specialize algorithms for each class. We call this new concept the automatic generation of specialized algorithms.

The idea of creating specialized algorithms for distinct classes of instances introduces a fresh perspective to the realm of algorithm generation. By clustering similar optimization problem instances using a clustering method based on specific features of instances, it's possible to design algorithms tailored for those clusters. Overfitting, typically a concern in many domains, is reframed here as an advantage. Rather than aiming for a universally applicable solution, this approach focuses on achieving the best solution for a narrower context. This concept challenges conventional views and can lead to higher efficiency and more effective solutions within distinct problem categories.

4.9 AGA FOR POPULATION-BASED ALGORITHMS

The solution container concept can be expanded to store a population of solutions. If terminals consider genetic operators like those typically used in evolutionary programming, the solutions to the master problem could give rise to many variations of genetic algorithms. By expanding the solution container concept to hold multiple solutions, we essentially allow for a meta-evolutionary process. This solution container extension paves the way for a system that searches for optimal solutions and simultaneously evolves the metaheuristic to find those solutions. This self-evolving structure might lead to more efficient and tailored algorithms for specific problems over iterations as the system "learns" the most effective operators and sequences of operations.

Allowing the solution container to manage multiple solutions simultaneously is a transformative leap in generating algorithms, particularly evolutionary algorithms like GAs. By not restricting the solution container to traditional GA representations such as float point, integer, binary, or permutations but instead allowing for their combinations or even entirely new representations, we open the door to a much broader search space for potential algorithms.

In traditional GAs, the representation of solutions often dictates the kind of genetic operators. By enabling multiple representations or hybrid representations in the solution container, the AGA framework can potentially discover novel crossover and mutation techniques that might be more suited to specific optimization problems than conventional methods. This flexibility can lead to the emergence of more adaptive, robust, and efficient algorithms than classical GAs. Furthermore, this approach aligns with the modern algorithmic design perspective, favoring adaptability and hybridization over rigid, singular methodologies. The ability to combine the strengths of various representations might lead to synergies previously unexplored. For instance, a solution representation that combines binary and permutation elements could allow for intricate encodings of solutions that benefit from both representations' advantages.

Expanding the Solution Container's flexibility to encompass multiple or hybrid representations would allow us to generate a new class of evolutionary algorithms, which could be more versatile and practical than traditional methods. This innovation holds great promise for advancing the field of metaheuristics and could redefine how we approach combinatorial optimization problems.

4.10 REDISCOVERING ALGORITHMS

Since the dawn of the computing age, many algorithms have emerged in academic literature. Remarkably, exact algorithms—those designed to find the optimal solution for every instance of an optimization problem—are grounded in intricate theoretical foundations. Iconic algorithms like Dijkstra, Simplex, and Branch and Bound have undergone extensive study and refinement. Their performance enhancements over recent decades can be attributed to advancements in computational power and profound theoretical advancements. The development of these algorithms often represented monumental efforts by researchers worldwide, spanning several decades in some cases. The foundational algorithms represent the culmination of human ingenuity, rigorous mathematical proofs, and iterative refinement. They have withstood the test of time, adapting and evolving as both computing capabilities and our theoretical understanding have expanded. From such development arises the question: could a modern AGA's framework replicate these classic algorithms, and if so, how long might such a process take?

Introducing the AGA framework into this narrative raises intriguing possibilities. Could such a framework, equipped with the right primitives and guided by the right objectives, eventually reproduce or even surpass these classical algorithms? On the one hand, it's tempting to think that with enough computational power and the proper set of building blocks, the framework might stumble upon similar structures or strategies. After all, if these classical algorithms represent optimal or near-optimal solutions to specific problems, then any sufficiently advanced search method might converge toward them.

If the automatic generation framework were to embark on a quest to rediscover these classics, it would not just be a matter of computational power. It would require a carefully curated environment rich in the right kind of primitives and knowledge and guided by objectives that can steer the search in meaningful directions. Even then, the timeline for such an endeavor remains uncertain and motivates future research.

4.11 AGA SPECIFICATION SHEET

We can use a Specification Sheet (SS) to specify all components needed to apply AGA for a specific optimization problem. The AGA SS is a streamlined and cohesive tool designed to facilitate the AGA through GP. Acting as a one-page blueprint, it briefly outlines all the foundational elements of a specific optimization problem. The sheet provides a comprehensive guide from a clear problem statement and relevant references to the functions, terminals, and crucial GP parameters. Its design promotes clarity, replicability, and standardization, ensuring that both novices and experts can quickly grasp the problem's intricacies and the GP process's nuances. By encapsulating the essence of the problem and the algorithm generation methodology in a concise format, the AGA SS stands as a pivotal tool for efficient research, communication, and innovation in combinatorial optimization.

Figure 4.7 presents an example. As revised in the last example, the SS contains all the information to apply AGA to the BKP. The sheet starts with a concise description of the optimization problem, providing a clear

AGA Specification Sheet

Optimization problem statement

Binary Knapsack problem (BKP). The binary knapsack problem consists of selecting a subset of items from a given set, each with an associated profit and weight, to maximize the total profit while adhering to a constraint on the total weight.

References

- Martello, S.; Toth, P. (1990). Knapsack Problems: Algorithms and Computer Implementations. Bologna, Italy: John Wiley & Sons.
- Pisinger, D. (2005) Where are the hard knapsack problems? Computers & Operations Research, 32(9), 2271–2284.

Solution container

A solution container for the BKP is a binary array of size n, corresponding to the total number of items available. Within this string, each position j signifies whether the respective item j is chosen for inclusion, represented by '1' or exclusion, indicated by '0' from the knapsack.

Set of functions

- *If-Then*: *If-Then*(P_1, P_2) executes operand P_2 only if operand P_1 returns True. The function returns True if P_1 returns True.
- *Or*: *Or*(P_1, P_2) executes operand P_2 only if P_1 returns False and returns True when P_1 or P_2 has returned True
- *And*: *And*(P_1, P_2) executes both operands and returns True if both operands also return True. Otherwise, it returns False.
- While: *While*(P_1, P_2) executes the terminal or function P_2 as long as P_1 continues to return True. Then, the process repeats until P_1 returns False. In addition, an upper limit to the running time or number of repetitions is necessary to avoid infinite operation.

Set of terminals

- T_1 (Add-Item): Add the next available item to the knapsack feasibly.
- T_2 (Remove-Item): Remove an item from the knapsack.
- T_3 (Add-Best-Value): Add the best profit-weight ratio item to the knapsack feasibly.
- T_4 (Remove-Worst-Valued): Removes the worst profit-weight ratio item

Master problem objective function

s_i^*: is the optimal value of the problem instance *i*. There are *m* instances.

s_i: the value found by an algorithm π.

f_π^{MSE}: is the objective function of the master problem. Mean squared error.

$$f_\pi^{MSE} = \frac{1}{m}\sum_{i=1}^{m}(s_i - s_i^*)^2$$

Set of instances

BKP1, BKP2,...,BKP800

GP parameters

Population size: 500; Crossover prob.: 0.8; Mutation prob.: 0.2. Random seed: 32.

FIGURE 4.7 AGA Specification Sheet for the binary knapsack problem.

context. Also, the SS includes relevant references, which can be helpful for anyone unfamiliar with the specific problem or looking for deeper insights. First, specify the solution container ensuring that generated algorithms produce valid outputs. Next, the solution space defines all available functions for algorithm construction, providing clarity on the logic that the GP can employ. These functions are generally helpful for many optimization problems, as they are not dependent on any particular one. Enumerate all available terminals or actions to ensure the GP has a specific set of operations it can perform to modify or interact with the solution. We also need a clear objective function for the master problem to guide the evaluation and selection of algorithms. Also, specify the set of instances, separating them into two groups: one for generating algorithms and the other for evaluating the final selected algorithms. In addition, specify the GP parameters necessary to run the machine, ensuring the GP process is replicable and tailored to the problem.

The AGA SS is a concise and comprehensive tool for automatic algorithm generation via GP. Consolidating all essential information onto a single page ensures clarity and simplicity. This standardization fosters replicability and consistent research outcomes and streamlines communication among research teams, ensuring everyone operates from a shared understanding. This tool offers a holistic perspective on the problem and the corresponding GP parameters, facilitating adaptability and promoting research consistency.

4.12 SUMMARY

This chapter introduces the concept of the AGA within the framework of GP, highlighting its potential to revolutionize algorithm design and optimization. It traces the evolution from Alan Turing's pioneering work to John Holland's theories on adaptation, setting the stage for John Koza's formalization of GP. The discussion delves into EC, a subset of artificial intelligence inspired by biological evolution, to develop algorithms that evolve through selection, mutation, and recombination operations. Central to this exploration is modeling the "Master Problem" using GP to discover computational strategies that yield optimal or near-optimal solutions. The chapter examines solution representation via tree structures, algorithm construction and refinement, and the balance between robustness and specialization. It further investigates AGA's capacity to generate population-based algorithms, integrate diverse metaheuristics, and rediscover classical algorithms. By dissecting and recombining elemental components of metaheuristics, AGA creates hybrid algorithms with enhanced efficiency and effectiveness. The chapter concludes that AGA, empowered by GP, offers a transformative approach to algorithmic design, enabling the automatic generation of specialized algorithms tailored to specific problem instances and raising intriguing questions about computational creativity and the automation of discovery.

EXERCISES

1. In the Job Shop Scheduling Problem, a set of jobs, each consisting of a sequence of tasks, must be scheduled on a set of machines to minimize the total completion time without any machine handling multiple tasks simultaneously. Generate a tree representation of an algorithm to find near-optimal solutions for this problem. Use internal nodes for operations (e.g., IF-THEN, WHILE) and leaf nodes for actions (e.g., schedule tasks, check machine availability).

2. Decompose the well-known Dijkstra and Floyd algorithms for the shortest path problem into their elementary components, such as initialization steps, distance updates, and path relaxation. Using these components, propose a new hybrid algorithm that leverages strengths from both algorithms. Create a syntax tree representing this new algorithm, where internal nodes are functions and leaf nodes are terminals with specific actions on the container solution.

3. Given a fleet of vehicles and a set of customers with known locations and demands, the goal is to determine the optimal set of routes for the vehicles to minimize the total distance traveled while ensuring all customer demands are met, each vehicle starts and ends at the depot, and vehicle capacities are not exceeded. Model this situation to the metaheuristic GP following the framework used in examples 4.1 and 4.2.

4. The Maximum Independent Set (MIS) problem is a classical combinatorial optimization problem where the objective is to find the largest subset of vertices in a given graph such that no two vertices in the subset are adjacent. Begin by defining the objective function (maximize the size of the independent set) and the constraints (no two vertices in the set are adjacent). Represent the space of possible algorithms, parameter settings, and problem instances as a multidimensional search space. Develop a GP framework where trees represent potential algorithms, with internal nodes as functions and leaf nodes as specific actions on the current solution. The goal is to evolve these trees to rediscover an efficient algorithm for

finding the maximum independent set, optimizing performance metrics such as the size of the independent set and computational efficiency. Present your proposed master problem formulation and the initial setup for the GP process.

5. Select an NP-hard optimization problem, such as the Traveling Salesman Problem (TSP), and formulate the master problem to be solved using a genetic algorithm (GA). Begin by defining the objective function and constraints specific to the selected problem. Describe how to represent the problem for the GA approach, including encoding solutions as chromosomes, defining the fitness function, and specifying genetic operators like selection, crossover, and mutation. Develop a GA framework to search the algorithmic space, optimizing the performance metrics such as solution quality and computational efficiency. Present your proposed master problem formulation and the initial setup for the GA process.

6. In the context of GP, constructive algorithms build solutions from scratch by adding components step by step, while refinement algorithms improve existing solutions by making incremental adjustments. Design a constructive and refined set of functions and terminals to compose algorithms for the Traveling Salesman Problem (TSP). Define the initial setup to solve the TSP, including the functions and terminals necessary for the GP framework to evolve efficient algorithms. This setup will help find near-optimal tours in TSP instances by leveraging the strengths of both constructive and refinement approaches.

7. Discuss the importance of parameter settings in GP. Identify all parameters involved and create a parameter tuning plan for a GP algorithm to solve the Knapsack Problem.

8. Define an objective function for the master problem that aims to search only for polynomial-time algorithms for the TSP. This function should evaluate the performance of candidate algorithms based on their ability to find near-optimal solutions within polynomial time complexity. Consider incorporating metrics such as the quality of the solution and the computational efficiency to ensure that the selected algorithms produce near-optimal solutions and do so efficiently within polynomial bounds. Present the formulation of this objective function and explain how it guides the search process toward identifying polynomial-time algorithms for the TSP.

CHAPTER 5

AGA and Machine Learning

5.1 INTRODUCTION

Optimization is a powerful tool to address problems arising in machine learning (ML) and automatic generation of algorithms (AGA). Optimization is a foundational concept in many scientific and engineering disciplines. It's about seeking the best solution from a domain containing possible candidates. In engineering, optimization is a field that allows modeling real-world situations employing a mathematical structure. In addition, it has the algorithms necessary to determine the optimal solution of such a structure. There are several mathematical models, such as linear programming, nonlinear programming, stochastic programming, integer programming, multiobjective programming, and several other variants. Then, the initial stage of engineering in this field consists of studying the problem's reality, specifying the problem's requirements, and then mapping those requirements onto the mathematical model. With that mathematical model, we select the appropriate algorithm to solve the real-world problem. AGA and ML consider optimization models to represent the phenomenon recorded in a dataset. In ML, determining the parameters of such mathematical models occurs through a minimization between the data yielded by the structure and the observed data.

DOI: 10.1201/9781003542896-5

The optimization process for solving the master problem described in Chapter 2 identifies an equation adjusted to the data in ML, which is about adjusting a mathematical relationship to a dataset. Given a set of inputs and corresponding outputs, the goal is to find a function that maps these inputs to outputs most accurately. A mathematical model, like a neural network or a linear equation, typically represents this relation. The model establishes a relationship between independent and dependent variables. The inputs and outputs are data of a given phenomenon recorded historically. The mathematical model identifies that relationship; however, to specify it completely, it is necessary to know its parameters. For instance, in a linear regression model, these parameters could be the slope and the intercept, and the final model is a linear equation with the dependent variable as a function of the independent variables. However, the challenge is determining the best values for these parameters. That's where optimization comes into play. The process begins by defining an objective function. Given a particular set of parameter values, this function quantifies how well the model's predictions align with the correct outcomes. A lower loss function value indicates a better fit of the model to the data. Different error measures, such as mean squared or absolute error, work well as an objective function.

AGA is an advanced concept that embodies the automatic assembly of atomic instructions composing the best algorithm for a specific problem. Here, optimization emerges as a pivotal player employing the master problem presented in Chapter 2. An optimization algorithm searches the vast space of potential algorithmic combinations and configurations to find the one that best addresses the specific optimization problem. The space of possible algorithms generated by various combinations of atomic components is vast, much like the high-dimensional parameter space of an ML model. Just as a loss function in ML measures how well a model's predictions match the actual outcomes, in AGA, a performance metric or evaluation function in the master problem is employed to determine how effectively a particular algorithmic configuration solves the specific optimization problem. Optimization techniques, such as genetic programming, genetic algorithms, simulated annealing, or gradient-based methods, can then be applied to explore this algorithmic space. They iteratively refine the combination of atomic algorithm or heuristic components, adjusting and reassembling them to improve the performance metric. The goal is to find an optimal algorithm capable of finding a near-optimal solution that performs significantly better than generic or traditionally hand-crafted algorithms.

In essence, AGA and ML rely on optimization as their driving force. While ML focuses on tailoring mathematical models to represent data patterns, AGA customizes algorithmic structures for specific computational challenges. In both cases, the magic lies in navigating a vast search space to find the best solution, making optimization an indispensable tool in modern computational problem-solving. Optimization forms the base, offering tools and methodologies for iterative improvement. Thus, AGA capitalizes on optimization to search the vast landscape of algorithms or model architectures.

5.2 SCHEMATIC OVERVIEW OF MACHINE LEARNING

Considers Figure 5.1 to have an overview of the ML field. The image shows on the left a representation of a general dataset consisting of various input values: $x_1, x_2, \ldots$, leading to an output y. This data is an input to the optimization problem, which aims to minimize a given error function represented by $E(w)$. This optimization task finds the parameters of a hypothesized mathematical model. The image contains several mathematical model candidates to adjust to the dataset. The models range from simple linear combinations of input variables to more complex equations involving exponential terms. Among these equations is a neural network at the bottom of the image represented as a graph. Each node corresponds to a specific term of the complex equation, and each arc represents an algebraic connection between the terms of the equation, and its weight is the proper parameter w. On the right of the image, the section showcases several resultant formulas derived from the optimization process that are the same formulas hypothesized but now with the parameter values already known. Thus, ML disposes a set of techniques to obtain the parameters of a given equation representing the phenomena behind the data. While Figure 5.1 provides a comprehensive overview of certain aspects of the ML field, it's important to note that it does not encompass the entire spectrum of ML techniques. For instance, this figure does not depict methods like Random Forest, a powerful ensemble learning technique. As such, while the figure offers valuable insights into specific ML methodologies, it should be viewed as a representation of a subset of the broader and diverse field of ML.

5.2.1 Modeling a Practical Problem

Modeling a problem is a fundamental step in data analysis and ML. It involves framing the problem in mathematical or computational terms, determining which variables play significant roles, and understanding the

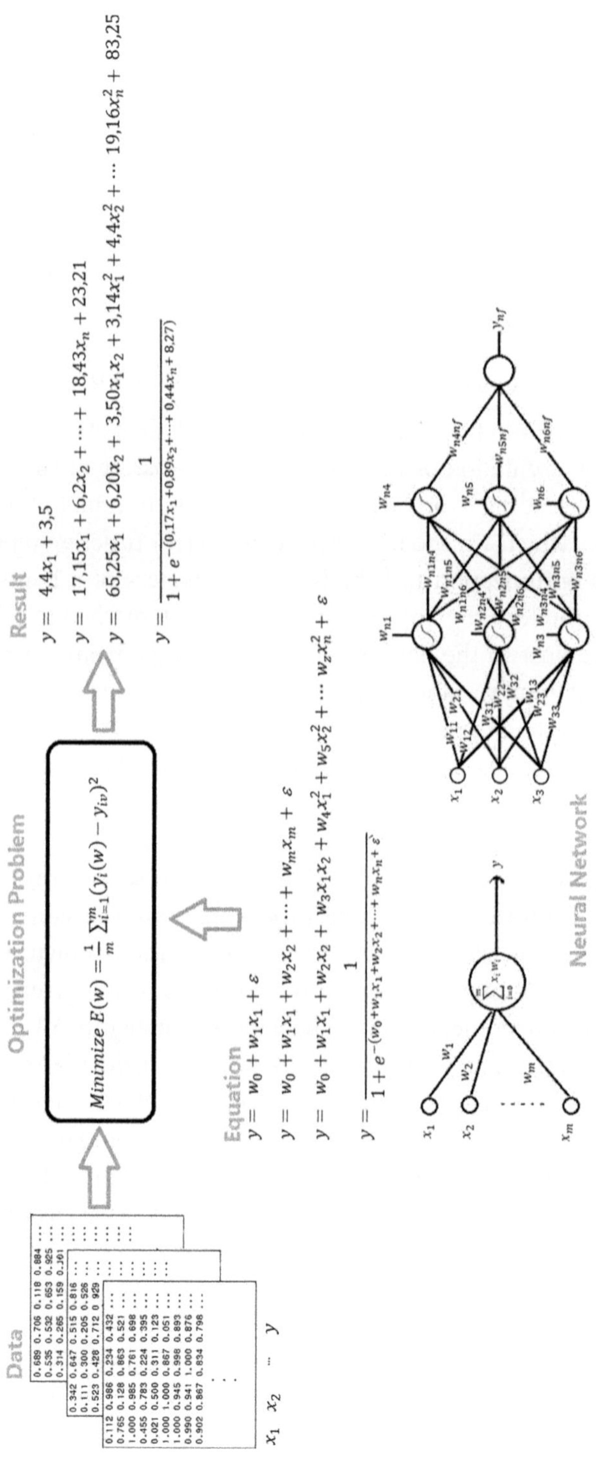

FIGURE 5.1 Schematic view of ML.

relationships between these variables. In Chapter 3, we revised this concept for real-world problem modeling. When we talk about modeling in the context of ML, we are essentially translating a real-world problem into a form that can be understood and acted upon by algorithms. At the heart of this translation is the identification of our variables. Modeling is not just about identifying these variables, though. It's also about understanding the relationships between them.

The variable of interest, often referred to as the target or dependent variable, is what we aim to predict or explain. It's the primary focus of our analysis. For instance, if we look at factors affecting a company's sales, the total monthly sales might be our variable of interest. The variables that help explain or predict our interest variable are independent variables or features. These are the inputs to our model. Using the same example, factors like advertising spending, product price, competitor activity, and seasonal trends might be our independent variables, as they can all influence sales and are independent among them.

Moreover, not all available variables are helpful or relevant in real-world scenarios. Part of the modeling process is the feature selection that identifies the essential variables for the analysis. Incorporating irrelevant features can lead to overfitting, where the model performs well on the training data but poorly on new, unseen data. The output of this step is a dataset shown in the Dataset Module of Figure 5.1. Chapter 3 presents a general framework to model real-world problems that we can apply to ML models. Then, the most common steps to approach a real-world problem with ML techniques are:

a) Identification of the Problem. Identifying a real-world problem requires a thorough understanding of the context and a deep analysis of the benefits and costs associated with potential solutions. One can pinpoint the core challenge by observing and listening to common issues. Evaluating the benefits of addressing the problem showcases its positive impact, such as improved efficiency, societal well-being, or economic gain. Conversely, understanding the costs, be they monetary, time, or resources, provides a realistic view of the feasibility of potential solutions. Engaging with stakeholders, from those affected to industry experts, ensures a holistic understanding. Documenting these insights, including the balance of benefits against costs, is crucial for clarity and building a persuasive case for action. This comprehensive approach ensures the problem is identified and ripe for effective solution development. In addition, reviewing scientific literature enables us to identify prior works and findings, placing our problem in a historical context and revealing gaps or opportunities for advancement.

b) Data Collection and Understanding. Gathering data relevant to the situation and understanding its features. Understanding the features of the data is equally crucial. Different data types have unique characteristics that can reveal specific facets of the situation. For instance, time-series data might expose trends over time, while categorical data can highlight distribution imbalances among groups. Moreover, the very nature of the data – its distribution, central tendencies, and variations – can offer clues about the underlying foundational problem. For example, if the data suggests a lot of variability in a process that should be consistent, it might indicate an issue with process control or external disruptions.

c) Feature Engineering. Transforming or constructing data features in a way that best represents the underlying patterns related to the objective. This process, often referred to as feature engineering, seeks to optimize how data interacts with algorithms or analytical methods, ensuring that the true essence of the information is captured and utilized. At its core, feature engineering revolves around understanding the nature of the data and the problem at hand. Different problems might require the data in distinct forms. For instance, while a time-series analysis might benefit from features highlighting trend components, a classification task might need features accentuating class separations.

d) Mapping to a Foundational Problem. It is advisable to map the problem to one of the foundational ML problems, such as regression, classification, clustering, or pattern recognition. This process requires understanding the situation, preparing the data, and mapping it to a specific ML problem. Mapping a real-world situation to one of the foundational problems provides a structured approach and allows leveraging existing algorithms and methodologies developed for these problems. Chapter 3 presents three of the foundational ML problems 3.

5.2.2 Meaning of the Dataset

The ML field is transversal because a dataset, composed of data captured from any particular phenomenon, serves as an element to find a mathematical model for the data and, consequently, for the phenomenon.

This universal applicability means that ML techniques cover various sectors and disciplines. From analyzing the nuances of human language in natural language processing, predicting stock market trends in finance, assisting in medical diagnoses, and even understanding consumer behavior in marketing, ML algorithms take in data, learn from it, and provide insights or predictions relevant to that field. This versatility has led to an interdisciplinary blend, where experts from diverse domains collaborate with data scientists to harness the power of ML, bridging the gap between domain-specific knowledge and advanced computational techniques. Consequently, the boundaries of what ML can achieve are continuously expanding, driven by the universality of data and the endless possibilities it presents when viewed through the lens of ML.

With more details, let's consider a few possibilities for variables x_1 to x_n in Figure 5.1:

- Medical Data: Each row could represent an individual patient, with columns reflecting various health metrics, like blood pressure, cholesterol levels, age, and weight. In the last column, y could indicate the presence (or absence) of a particular disease or health outcome.
- Financial Transactions: Each row might represent a transaction in a financial context. Columns could signify attributes like transaction amount, time of day, account balance before the transaction, etc. The y column could represent the likelihood of the transaction being fraudulent.
- Real Estate Listings: Each row might correspond to a property listing for a real estate dataset. Features could encompass the number of bedrooms, total area, proximity to amenities, building age, etc. The y column could denote the final selling price of the property.
- E-Commerce: In the context of online shopping, each row could depict an individual product. Columns might capture attributes like product weight, dimensions, average customer review score, and shipping time. The y column might represent the monthly sales volume of that product.
- Environmental Studies: Each row could denote daily recordings if studying environmental patterns. Columns could represent daily average temperature, humidity, wind speed, and pollution levels. The y column might symbolize the likelihood of rainfall on that day.

5.2.3 Hypothetical Model

The hypothetical model establishes a mathematical relation that equates the phenomenon variables. Such a model is a hypothesis that points out that this equation relates the dependent and independent variables. This model establishes a specific relationship between dependent and independent variables and is an initial conjecture in ML and statistics. It's a roadmap that hints at how changes in independent variables might influence the dependent variable, captured succinctly in the equation. However, while the model offers a structured perspective, its true merit is determined by its alignment with reality; theoretical constructs must withstand empirical scrutiny to gain acceptance.

A problem arises because the hypothesized equation has parameters whose values are unknown. The main objective of ML is to try to calculate such equation parameters. On the other hand, if the parameters are known, we have an equation that obeys the phenomenon registered in the data, i.e., an equation for the data. In this way, the dataset is a picture of the phenomenon. If the data ideally register the phenomenon, then the equation is the phenomenon equation. Having the equation wholly known, one can predict the unknown dependent variable regarding the feature values. Thus, once estimated, these parameters transform the general equation into a specific model tailored to the data at hand. For instance, the equation $y = w_1x + w_2$ has two unknown parameters: the slope w_1 and the intercept w_2. A linear equation is one where the relationship between the independent and dependent variables is linear. An optimization process involves adjusting w_1 and w_2 to minimize the difference between the predicted values of y and the observed values in the dataset. Linear regression is the most common technique for analyzing and predicting outcomes based on such relationships. It assumes a straight-line relationship between the inputs and the output. For instance, predicting sales based on advertising spend, assuming each dollar spent on advertising consistently impacts sales by a certain amount, would be a linear problem.

On the other hand, nonlinear equations capture more complex relationships between variables. Such relationships aren't strictly proportional and can vary in nature. As the complexity increases, simple linear models might no longer suffice, and more intricate modeling techniques become necessary. Neural networks are a prime example of this. They can model complex nonlinear relationships using layers of interconnected nodes (or "neurons"). Each connection has a weight, and each neuron applies a nonlinear activation function to its input, allowing the network to capture and represent nonlinear patterns in the data.

In many real-world scenarios, the relationships between variables aren't straightforward. Simple linear models might be computationally efficient and easy to interpret but can fall short in capturing the intricacies of the data. In contrast, nonlinear models, like neural networks, can capture these nuances and yield more accurate predictions but at the cost of computational complexity and reduced interpretability.

Choosing between linear and nonlinear models often involves a trade-off between simplicity and accuracy. The nature of the data influences the decision, the underlying relationships among variables, the computational resources available, and the specific goals of the analysis. Ultimately, whether one opts for a simple linear regression or a more complex neural network, the objective remains to accurately model and predict the variable or set of variables of interest based on the available data. The "Equation" in the Figure symbolizes this mathematical representation, serving as the foundation for subsequent analysis and prediction in ML.

5.2.4 The Optimization Problem

How do we determine the parameter values of a hypothesized equation? An optimization problem is formulated in ML to obtain the equation parameter values. Determining the parameter values of a hypothesized equation is central to ML. An optimization problem allows us to find the best set of parameters. The foundation of this optimization problem is the objective function that measures the error between the predicted values, as given by the model with a particular set of parameters, and the observed values in the dataset. The objective is to adjust the parameters to minimize the value of this objective function. For example, the most commonly objective function in linear regression is the mean squared error (MSE). The MSE calculates the average squared difference between the predicted and observed values for a given set of parameters. The optimization problem then becomes finding the parameter values that minimize the MSE.

Several optimization algorithms are capable of solving such an optimization problem. However, the optimization landscape can be challenging. There might be many local minima, saddle points, or flat regions. Advanced variants of gradient descent, such as stochastic gradient descent (SGD) and mini-batch gradient descent, along with algorithms like Adam or RMSprop, have been developed to navigate these challenges more efficiently.

The field of ML has borrowed concepts from classical optimization and statistics, often giving them new terminology more attuned to its paradigms. These renamed terms emphasize the unique aspects and challenges of ML, helping practitioners communicate more effectively within the context of the field. However, at their core, the foundational concepts from optimization remain integral, guiding the development and application of ML techniques. Here are some of the most common ML terms and their counterparts in the optimization field:

- Training: It is the iterative process of the optimization algorithm.
- Loss function: Objective Function.
- Weights (especially in neural networks) or coefficients: Model parameters.
- Regularization: Use of penalties function in the objective function.
- Learning Rate: A hyperparameter in gradient-based optimization algorithms.
- Epochs: Number of repetitions of the optimization algorithm.
- Batch: Subset of the data used in the algorithmic iterative process.

5.2.5 Algorithms for the Optimization Problem

The ML field is deeply connected to nonlinear programming due to the predominance of nonlinear loss functions, especially in advanced models like neural networks. Nonlinear programming, concerned with optimizing a nonlinear objective function subject to nonlinear constraints, offers a rich set of tools and techniques that have been adapted and extended for ML tasks. The quest to find optimal solutions in potentially complex landscapes is at the core of ML and nonlinear programming. The nature of nonlinear functions means their optimization landscapes could have multiple local minima and maxima, saddle points, and other challenges. Traditional linear programming methods falter in these terrains, necessitating specialized techniques.

Several optimization algorithms from nonlinear programming have been pivotal in ML:

- Gradient Descent and its Variants: At its heart, gradient descent is a search algorithm heavily used in nonlinear optimization. In the context of ML, especially with neural networks, the error surface can be highly nonlinear. Gradient descent and its more sophisticated variants, like stochastic gradient descent (SGD), mini-batch gradient descent, and methods like Adam and RMSprop, navigate this complex landscape to adjust model parameters iteratively.

- Conjugate Gradient Method: This method works by iteratively optimizing the objective function along conjugate directions, which are linearly independent and respect a specific conjugacy condition relative to the function's Hessian matrix. In each iteration, it moves in the direction of the steepest descent that's conjugate to all previous directions.
- Newton Method: This technique minimizes the objective function by iteratively updating the equation parameters. It leverages the second-order Taylor series expansion, using both the function's gradient and the Hessian matrix to find a point where the gradient is zero, corresponding to a local minimum or maximum of the function.
- Quasi-Newton Methods: These methods approximate the Hessian matrix (second-order derivative) to inform the search direction. These have been applied in deep learning and other ML algorithms to accelerate convergence.
- Penalty and Augmented Lagrangian Methods: Regularization in ML, like L1 and L2, corresponds to the penalty method from nonlinear programming. These methods add penalties for model complexity to the objective function to ensure more robust and generalizable models.

It's also noteworthy that as ML models become more intricate and as the volume of data grows, there's been a surge in the development of novel optimization techniques explicitly tailored to these challenges. Thus, the union between ML and nonlinear programming is natural and productive because ML benefits from the theoretical foundations and algorithms of nonlinear programming; in return, the novel challenges posed by ML push the boundaries of optimization, leading to the evolution and expansion of nonlinear programming techniques, combinations with heuristic methods, and automatically generating combinations of various methods.

However, while finding these optimal parameter values is crucial, it's also essential to ensure that the model doesn't become too tailored to the training data, a phenomenon known as overfitting. An overfitted model might perform exceptionally well on its training data but fail to generalize to new, unseen data. Thus, part of the ML challenge is finding parameters that minimize training error and ensuring that the model remains robust and generalizable across diverse datasets.

Metaheuristic methods can also solve the ML optimization problem. Metaheuristic methods are high-level procedures designed to find, generate, or select a heuristic that may provide a sufficiently good solution to an optimization problem, especially with incomplete or imperfect information or limited computation capacity. These methods have gained prominence due to their versatility and ability to find high-quality solutions in reasonable time frames for complex optimization problems. Certain metaheuristics have gained prominence in the context of ML due to their efficacy in handling high-dimensional spaces, non-convex loss landscapes, and complex optimization scenarios. In ML, complex matrix equations corresponding to neural networks are often derivable, allowing for the computation of gradients and using gradient-based optimization methods. However, metaheuristics can be crucial when the equations are not derivable or the optimization landscape is particularly challenging. Here are some of the metaheuristic methods commonly used in ML:

- Genetic Algorithms (GA): These methods are helpful in ML for feature selection, hyperparameter tuning, and even for evolving neural network architectures.
- Particle Swarm Optimization(PSO): It helps train neural networks as an alternative to traditional optimization techniques like gradient descent. It also helps in feature selection and hyperparameter tuning.
- Simulated Annealing (SA): SA is versatile and works for various ML problems, including neural network weight optimization and feature selection.
- Ant Colony Optimization (ACO): ACO applies for feature selection, clustering, and rule-based classifiers.
- Tabu Search: Particularly effective for combinatorial optimization problems, Tabu Search has been applied to feature selection, clustering, and other optimization tasks in ML.

- Differential Evolution (DE): DE is appropriate for continuous optimization problems and works in ML tasks like neural network training, hyperparameter tuning, and feature selection.
- Harmony Search: Occasionally used for neural network training, feature selection, and other optimization tasks in ML.
- Cuckoo Search: This technique has been used in neural network training, especially in determining the optimal weights and other optimization tasks.
- Firefly Algorithm: This algorithm has been applied to clustering, feature selection, and neural network training in ML.
- Bat Algorithm: Inspired by the echolocation behavior of microbats, it's used in various ML tasks, including feature selection and neural network training.

It's important to note that while these metaheuristics are potent tools in the ML toolkit, their effectiveness is problem-dependent. Choosing a particular metaheuristic over another often depends on the nature of the optimization landscape, the problem's constraints, and the desired trade-off between exploration and exploitation. Additionally, with the increasing computational power and the development of specialized optimization techniques for ML, some traditional algorithms have seen adaptations and hybrids, combining multiple metaheuristics' strengths.

The essence of ML revolves around refining these unknown parameters to transform a general hypothesis into a data-driven model. It's a balance between fitting the data closely enough to capture underlying patterns while remaining sufficiently general to be applicable beyond the training dataset. Thus, the essence of numerical experiments comes into play with numerical hypotheses undergoing rigorous validation using data. We can measure the model's predictive prowess by training it on a subset of data and evaluating its performance on unseen data. Metrics derived from this process offer a quantifiable gauge of the model's accuracy, reliability, and generalizability. If the model aligns well with the test data, it's a testament to the hypothesis's strength. Conversely, discrepancies hint at potential oversights or simplifications in the model, necessitating revisions. This iterative hypothesis, validation, and refinement process ensures that our mathematical representations are theoretically robust and empirically valid, bridging the gap between abstract conjecture and tangible reality.

5.3 TYPES OF PROBLEMS IN MACHINE LEARNING

The ML field provides some fundamental theoretical problems to represent real-world situations. ML, at its core, is about developing algorithms that can learn patterns from data. These fundamental theoretical problems are foundational frameworks in ML that correspond to many real-world situations. Chapter 3 includes three of these problems. These foundational problems provide the basic structure and methodology for building ML solutions. When faced with a real-world situation, one of the initial tasks is to determine which of these foundational problems the situation aligns with most closely. By doing so, we can leverage the vast array of algorithms and techniques specifically developed for these problems to address the given situation. Chapter 3 presents a general framework for modeling real-world situations. Here is a brief description of some of them, which are functional to represent many real-world situations.

- Regression problem: This problem helps predict a continuous value. For example, predicting house prices based on features like size, location, and number of bedrooms is a regression problem.
- Classification problem: This problem involves categorizing data into predefined classes. Diagnosing diseases (diseased or healthy) based on patient symptoms or medical test results is an example of a binary classification problem.
- Clustering problem: Unlike classification, where classes are predefined, clustering involves grouping data based on their similarity without having prior labels for the groups. It's helpful in exploratory data analysis to find the data's inherent structure. An example is customer segmentation, which identifies different groups based on purchasing behavior.
- Reinforcement Learning problem: This is a type of problem where an agent learns to make decisions by taking actions in an environment to maximize a cumulative reward. Its focus is on learning from interaction and exploration, and its use of feedback in the form of rewards (or penalties) to guide this learning. This theoretical problem helps represent situations like training algorithms to play and often excel in complex games or optimize advertisement delivery.
- Time Series analysis and forecasting problem: Time series analysis involves understanding and extracting meaningful statistics and

characteristics from a sequence of observations ordered in time. The primary goal is to describe and model the underlying structure in the data. In turn, time series forecasting involves using models to predict future values based on previously observed values. The aim is to capture the underlying patterns in the historical data to project these patterns into the future. Typical examples are stock market predictions, weather forecasting, and sales forecasting.

- Pattern recognition problem (PR): This field broadly refers to studying and designing computational models that recognize patterns, especially in data. PR encompasses various tasks, such as classification, clustering, and regression. The primary goal is to identify regularities, structures, or recurring patterns in data and then make decisions based on these recognitions. One example is anomaly detection in credit card use, whose goal is to identify data points that deviate from the expected pattern or behavior, in this case, to identify frauds. Anomaly detection explicitly looks for irregularities or patterns that don't conform to expected behavior.
- Recommendation problem: It aims to predict a user's preference or rating for an item or suggest items that might interest a user. These systems have become ubiquitous and are integral to online platforms, be it e-commerce, online media outlets, social media, or various other industries. Real-world applications recommend movies, suggest products on e-commerce sites, or provide song playlists on music platforms.

The foundational and integrated problems described above offer a comprehensive toolbox for representing and solving many real-world situations using ML. The art and science lie in recognizing which foundational problem or integrated problem best maps to the given real-world scenario. These problems provide the theoretical underpinning for many ML applications. Researchers and practitioners can leverage various algorithms and methodologies developed for these foundational problems to address practical challenges by translating real-world situations into one of these frameworks.

5.4 SCHEMATIC OVERVIEW OF THE AUTOMATIC GENERATION OF ALGORITHMS

Analogous to Figure 5.1, Figure 5.2 depicts a diagram highlighting the concept of the AGA applied to combinatorial optimization problems. The process considers three main components. The data component

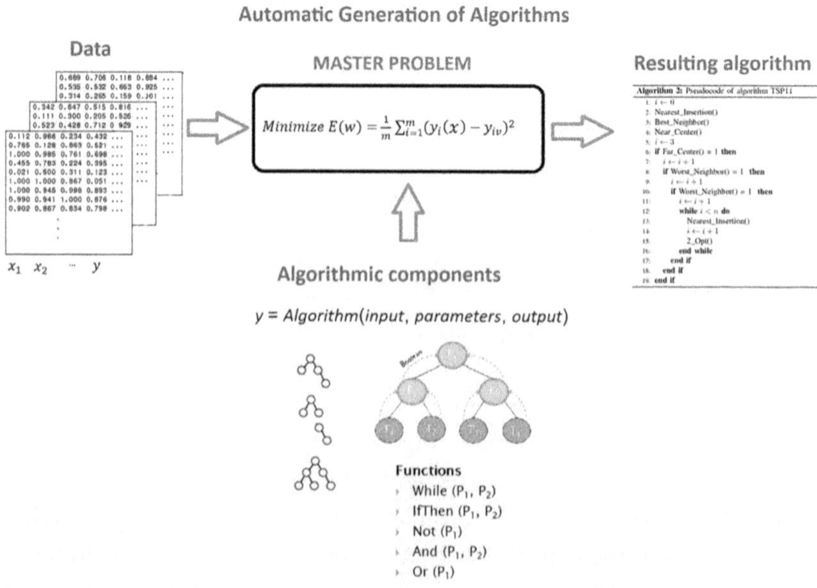

FIGURE 5.2 Schematic view of AGA.

corresponds to instances of specific optimization problems. Then, what in Figure 5.1 was the hypothesis equation is now elemental algorithmic components. The final algorithm is on the right side of the figure.

5.4.1 Problem Instances

When utilizing AGA to solve combinatorial optimization problems, the known optimal solutions for these instances serve as a benchmark. They enable the assessment of the algorithm's accuracy and efficiency. In practical scenarios, while it's computationally challenging to determine the exact optimal solution for large instances, having optimal solutions for smaller instances aids in training models and gauging the effectiveness of new heuristic or metaheuristic approaches. Consider the case of the Traveling Salesman Problem as an example. For that case, the "Data" would represent multiple instances of cities to be visited. Each instance would detail the cities and the distances between them. For practical training and validation in an AGA context, it's crucial that each instance also has an associated optimal solution.

Splitting the data into training and validation sets is a standard approach in ML. The same splitting is necessary in AGA to produce algorithms for optimization problems like the Traveling Salesman Problem. The training set allows us to construct the algorithm for the optimization problem.

The iterative training process allows the algorithm to adjust its parameters and strategies based on feedback from the known optimal solutions for the training instances. Over time, this refines the algorithm's performance. In addition, testing its effectiveness on unseen problem instances is essential. The validation set provides this new set of challenges for the algorithm, and its performance on this set gives insights into how well the algorithm might perform on entirely new problem instances or even in real-world scenarios. By assessing performance on the validation set, one can gauge the algorithm's generalization capability, ensuring that it's not just memorizing the training instances but genuinely learning to solve the problem.

In the context of the TSP, the training instances might involve specific patterns of cities, distances, or configurations, while the validation instances might introduce new patterns or challenges. By effectively training on one set and validating on another, one ensures the constructed algorithm is robust, adaptable, and effective across a broad spectrum of TSP instances. This approach mirrors the train-test split used in traditional ML tasks, underscoring the convergence of optimization and ML methodologies.

5.4.2 Algorithmic Components

The algorithmic components are a crucial part of the AGA process. It employs a tree-like structure corresponding to a syntax tree to represent and assemble the algorithm. This tree-based representation is a hierarchical structure where each node denotes a specific function or operation. The tree encapsulates the algorithm's computational logic, with the operations flow moving from the root down to the leaves. The internal nodes represent functions that are standard algorithmic operations like loops, conditionals, or logical operators. These functions dictate the flow and control of the algorithm, determining the subsequent operations or decisions. The syntax tree's leaf nodes, or the terminal nodes, are the actual operational components. In the context of combinatorial optimization, these terminals operate on the container of the current solution. The operations at these terminals directly manipulate or evaluate parts of the solution space, driving the algorithm toward an optimal or near-optimal solution. As the tree structure gets executed, it dynamically interacts with the current state of the combinatorial optimization problem solution. This approach harnesses the power of tree-based representations, commonly seen in genetic programming and other heuristic methods, as presented in Chapter 4.

5.4.3 Master Problem

The master problem in Figure 5.2 corresponds to a higher-level optimization problem focusing on optimizing the optimization process presented in Chapter 2. Thus, the master problem concerns discovering the most effective algorithm to solve a primary optimization problem. The primary aim is to find an algorithm that can yield a near-optimal solution for the original combinatorial optimization problem. This scheme is distinct from directly finding the solution to the problem; instead, it's about finding the best mechanism, an algorithm, to get to that solution. A performance measure reflects the derived algorithm's quality, representing its computational efficiency when applied to the primary optimization problem. It essentially quantifies the discrepancy between the solution provided by the algorithm and the known optimal solution. The sought algorithm should not only be effective but also human-readable. This requirement imposes a size or complexity constraint, ensuring that the resulting algorithm is interpretable and not overly intricate. In this way, the search process occurs in the space encompassing all potential algorithms that could address the primary optimization problem. The boundaries of this space are determined based on the desired attributes of the target algorithm.

5.4.4 The Resulting Algorithm

Transforming a syntax tree into a pseudocode involves traversing it, converting each node and its associated branches into equivalent pseudocode statements. The process typically depends on the structure and nature of the syntax tree and is represented as the resulting algorithm in Figure 5.2. For each node encountered, If the node represents a function or terminal, translate it to the corresponding pseudocode statement. For example, a node representing addition might translate to an "ADD" operation in pseudocode. Otherwise, if the node is a function, translate it into the corresponding control statement in pseudocode, ensuring that any child nodes or branches are included. If the node is a terminal, it might represent inputs, outputs, or fixed values. In this case, represent it as the appropriate variable or value in the pseudocode. If a node has multiple child nodes, ensure the pseudocode properly represents their relationship. This task might involve using logical operators like "AND" or "OR" for decision-making nodes, defining loops for nodes representing iterative processes, or nesting control structures if one child node is another control structure. Continue the traversal and translation process until all tree nodes are processed.

5.5 SYMBOLIC REGRESSION

Symbolic regression closely aligns with the concept of AGA and serves as a bridge between ML and AGA. In AGA, genetic programming evolves algorithms that solve specific problems, as described in Chapter 4. In the case of symbolic regression, GP works to generate mathematical expressions that fit the data automatically. Symbolic regression is a form of regression analysis that searches for mathematical expressions, represented as symbolic formulas, to best fit a given dataset. Unlike traditional regression, which fits data to a predetermined model structure, symbolic regression aims to discover both the model structure and the parameters simultaneously. This task occurs by exploring a vast space of possible mathematical expressions and selecting the one that best describes the underlying relationship in the data. The evolutionary process iteratively refines these expressions, akin to how AGA develops and optimizes algorithms. Symbolic regression thus exemplifies AGA by automatically deriving the optimal mathematical model for a given dataset.

The symbolic regression problem can be formally defined as follows: Given a dataset $D=\{(x_i, y_i)\}_{\{i=1\}}^{N}$, where x_i represents the input variables and y_i represents the corresponding target values, the objective is to find a mathematical expression $f(x)$ that minimizes the prediction error. The objective function is typically the MSE between the predicted values $\hat{y}_i = f(x_i)$ and the target values y_i:

$$\text{Minimize} \quad \frac{1}{N}\sum_{i=1}^{N}\left(\hat{y}_i - y_i\right)^2$$

The constraints involved in symbolic regression include:

- **Complexity Constraint**: Limit the complexity of the expression to avoid overfitting, often measured by the syntax tree's depth or number of nodes.
- **Function and Terminal Validity**: Ensure that all functions and terminals used in the expression are valid and well-defined.
- **Domain Constraints**: If applicable, ensure the expression adheres to domain-specific constraints, such as non-negativity or bounded outputs.

The mathematical expressions correspond to syntax trees, where the internal nodes are functions such as arithmetic operations or trigonometric functions, and the leaf nodes correspond to terminals that are variables and constants. The genetic programming process involves generating an initial population of random syntax trees, evaluating their fitness based on the objective function, and through iterative refinement, symbolic regression discovers the optimal mathematical model that best captures the underlying relationship in the data.

Neural networks and symbolic regression represent two distinct approaches to modeling and understanding data within the ML landscape. Neural networks, with their structurally fixed architectures, rely on layers of neurons connected through weighted edges. The relationships between inputs and outputs in neural networks correspond to complex algebraic or matrix equations embedded within the network's architecture. The training process optimizes these equations by adjusting the weights and biases of the network to minimize an error function. In contrast, symbolic regression employs a more flexible scheme to discover the mathematical expressions that best represent the data. Instead of being constrained to a fixed structure, symbolic regression explores a vast space of potential expressions, using genetic programming to iteratively refine these expressions based on their fit to the data.

This parallel between neural networks and symbolic regression is particularly intriguing given the historical context of ML. Neural networks have been extensively explored and applied across various problems, often becoming the go-to approach for tasks such as image recognition, natural language processing, and more. Despite this, symbolic regression offers apparent advantages in flexibility. Symbolic regression not only optimizes the parameters within an expression but also discovers the structure of the expression itself, providing a more holistic and potentially insightful data model. In neural networks, the optimization process focuses solely on adjusting the parameters of a predefined structure. As a result, the final model consists of known parameters embedded within a fixed algebraic or matrix equation.

This distinction highlights the unique capabilities of symbolic regression in uncovering new and potentially simpler representations of complex phenomena. While neural networks excel in handling large, high-dimensional datasets with intricate patterns, symbolic regression provides an alternative for discovering interpretable models that reveal the fundamental mathematical relationships within the data. As ML continues to evolve, the complementary strengths of neural networks and symbolic regression offer valuable tools for tackling various challenges, underscoring the importance of flexibility and interpretability in developing robust and practical models.

Furthermore, the AGA framework allows us to increase the flexibility of modeling the dataset. Instead of trying to fit even complex equations to the data, the AGA framework allows for creating highly adaptable structures that can represent the data in algorithms rather than static equations. This approach provides an even greater degree of flexibility compared to both symbolic regression and neural networks. AGA evolves algorithms through genetic programming, enabling the discovery of innovative and tailored solutions that can adapt to the specific characteristics of the data. In this sense, AGA is more flexible than symbolic regression and neural networks in representing data in ML, as it can dynamically generate algorithms that optimize the problem-solving process based on the data's unique features. This adaptability makes AGA a powerful tool for developing robust and efficient models across diverse and complex ML tasks.

Example 5.1

Given a dataset containing information about house prices, with features such as square footage, number of bedrooms, number of bathrooms, and location. Your task is to predict house prices using three techniques: Neural Networks, Symbolic Regression, and AGA.

Consider the following steps to find a solution:

a) Neural Networks Approach:
   - Define the Architecture: Input layer with nodes for each feature. Several hidden layers with a specified number of neurons. Output layer predicting house prices.
   - Preprocess the Data: Data cleaning. Normalize features. Split the dataset into training and test sets, for example, 80% and 20%.
   - Train the Neural Network: Use a suitable optimization algorithm to minimize the MSE. Train the model on the training set and evaluate it on the test set.
   - Evaluate Performance: Calculate performance metrics such as MSE, RMSE, and R^2 score on the test set.

b) Symbolic Regression Approach:
   - Set Up Genetic Programming: Define the terminal set (square footage, number of bedrooms, bathrooms, location, constants). Define the function set, including addition, subtraction, multiplication, and division.

  - Evolve Mathematical Expressions: Generate an initial population of random expressions. Evaluate each expression based on its fit to the training data using MSE. Apply genetic operations to evolve expressions over multiple generations.
  - Select Best Expression: Choose the expression with the lowest error on the validation set.
  - Evaluate Performance: Test the selected expression on the test set and calculate performance metrics.

c) AGA Approach:
  - Define Components: Develop the formulation of the Master Problem. Define the sets of functions and terminals, the objective function, and the solution container. Define the initial population of algorithms.
  - Evolve Algorithms: Evaluate the performance of each algorithm on the training set using MSE. Apply genetic operations to evolve the population over multiple generations.
  - Refine and Select the Best Algorithm: Select the best algorithm on the validation set.
  - Evaluate Performance: Test the selected algorithm on the test set and calculate performance metrics.

Example 5.2

Consider a dataset containing information about a company's employee performance ratings. The features include years of experience, education level, number of completed projects, and department. We are required to predict the performance rating of employees using the AGA framework. The goal is to evolve an algorithm that generates an equation or a set of equations to predict the performance rating in a specific order, following the logic of operations defined in the algorithm.

We have to define the AGA elementary components.

- Solution Container: The solution container will store the weights of the different equations that play a role in the resulting algorithm. For example, suppose we have a linear equation with n_1 weights, a neural network with n_2 weights and a polynomial equation with n_3 the container will then store all these weights and allow for their updating during optimization.

- Terminal Set: Terminals are operations that update the weights and use the features to generate predictions.
  - Linear Equation: $y = w_{l0} + w_{l1}x_1 + w_{l2}x_2 + \ldots + w_{\text{ln}}x_n$
  - Neural Network: Predicts y using a neural network with weights.
  - Polynomial Equation: $y = w_{p0} + w_{p1}x_1 + w_{p2}x_1^2 + w_{p3}x_2 + \ldots + w_{pn}x_n^2$
  - Update Weights: A function to update the weights based on gradient descent or another optimization method.
  - Gradient Method: Applies a gradient-based optimization method to update the weights.
  - RandomSelectFeatureIndex: Randomly selects a feature index from the available features.
  - RandomSelectThreshold: Randomly selects a threshold value.
- Function Set: Functions are the logic operations that control the flow of the algorithm and combine terminals to form the overall prediction logic, as described in Chapter 4. Consider the following functions:
  - While-loop: Repeats a set of operations until a condition occurs.
  - If-then-else: Executes one operation if a condition is True and another if it is False.
  - And: Combines two conditions and proceeds if both are True.
  - Or: Combines two conditions and proceeds if at least one is True.
  - For-loop: Iterates over a set of operations a specified number of times.
  - ConditionLessThan(value1, value2): Checks if value1 is less than value2. Applicable when both variables have numerical values.
- Construct Initial Population: Generate an initial population of random algorithms using the defined functions and terminals. Each algorithm should be a tree structure with functions as internal nodes and terminals as leaf nodes.
- Fitness Function: The fitness function will evaluate the performance of each algorithm based on its prediction

accuracy. Use the MSE between the predicted and observed performance ratings as the fitness function.

$$\text{MSE} = \frac{1}{N}\sum_{i=1}^{N}\left(\hat{y}_i - y_i\right)^2$$

- Genetic Operations: Use GP to evolve the population of algorithms.
- Iterate and Select the Best Algorithm: Iterate over multiple generations, applying the fitness function and genetic operations to evolve the population. Select the best-performing algorithm based on the fitness function.

We can obtain a final algorithm like Algorithm 5.1 that optimizes the weights of multiple predictive models using a dataset. Initially, the algorithm defines essential parameters such as threshold, learning rate, maximum iterations, and initial weights for linear, neural network, and polynomial models. It iterates over the dataset for a specified number of iterations. For each data instance, it extracts the features (x) and the value (y), initializing the predicted value (y_pred) to zero. Depending on whether a specific feature ($x[3]$) is below the threshold, it either combines predictions from a linear equation and a neural network or uses a polynomial equation to determine y_pred. The algorithm then stores y_pred and computes gradients for each model, updating their respective weights to minimize prediction error, thereby refining the models iteratively.

Algorithm 5.1

```
BEGIN
  Input: dataset;
  Output: final weights for all equations;
  Define initial parameters: threshold, learning_
    rate, max_iterations;
  Define initial weights: weights_linear, weights_
    nn, weights_poly
  For i = 0 To max_iterations Do
   For each instance In dataset Do
    Select x and y;
    y_pred = 0
    If x[3] < threshold Then
```

```
        y_pred_linear = LinearEquation(x)
        y_pred_nn = NeuralNetwork(x,)
        y_pred = 0.5*y_pred_linear + 0.5*y_pred_nn
      Else y_pred = PolynomialEquation(x)
      Store y_pred
      gradients_linear = compute_gradients_linear
        (x, y_true, y_pred)
      update weights_linear;
      gradients_nn = compute_gradients_nn(x, y_true,
        y_pred)
      update weights_nn;
      gradients_poly = compute_gradients_poly(x,
        y_true, y_pred)
      update weights_poly;
END
```

5.6 SUMMARY

This chapter explores the relationship between AGA and ML, highlighting optimization as a pivotal concept bridging both domains. In ML, optimization facilitates fine-tuning mathematical models to datasets, aiming to minimize the discrepancy between predicted outcomes and observed data through parameter adjustment. AGA extends this principle to algorithm generation, employing meta-optimization to devise algorithms optimized for specific problem instances. This process comprehensively explores algorithmic components and configurations, guided by performance metrics like mean square or absolute value error, against a backdrop of known optimal solutions. Despite the shared optimization foundation, the objectives diverge: ML seeks broad applicability across diverse datasets, while AGA aims for peak performance in narrowly defined scenarios. This exploration underscores the multifaceted roles of optimization in computational problem-solving, bridging theoretical concepts with practical applications in algorithm generation.

Exploring AGA and ML underscores optimization's critical role in navigating the vast search space of potential solutions and algorithmic configurations. This chapter illustrates the complementary nature of AGA and ML, both rooted in optimization yet diverging in their pursuit of generalization versus specialization. It offers profound insights into the adaptable strategies essential for modern computational problem-solving.

EXERCISES

1. Model an optimization problem aiming to improve the performance of an ML model considering the parameters in the objective function. Use appropriate penalty functions. Define the objective function and constraints for this optimization problem, and explain how to use different optimization techniques like gradient descent or genetic algorithms to solve it.

2. Formulate the Master Problem for rediscovering classical ML algorithms like decision trees, k-nearest neighbors, or support vector machines using genetic programming. Define the search space, objective function, and constraints. Explain how the genetic programming framework would explore this space to identify small algorithms.

3. Define how to use the AGA technique to solve a classification problem based on a multilayer perceptron. Consider explicitly the classifying handwritten digits problem and represent the solution as an MLP model with input, hidden, and output layers. Use a solution container to store the MLP model's weights, biases, and activation functions. Define functions and terminals to build and modify the MLP. Create an objective function to evaluate the MLP's performance, balancing classification accuracy and model complexity. Implement a genetic programming framework to explore the search space of possible MLP models.

4. Model a hybrid optimization framework that combines elements of genetic algorithms and simulated annealing to solve the feature selection problem. Define the components of the hybrid model, including how the two optimization techniques will interact, the objective function, and the termination criteria.

5. Formulate constraints for the AGA that ensure the generated algorithms are practical, interpretable, and efficient. Explain how these constraints can be incorporated into the optimization process considering penalty functions.

6. Analyze the concept of overtraining in AGA versus ML. Begin by defining overtraining (or overfitting) in both contexts. Discuss the mechanisms that lead to controlling the overtraining in each approach.

7. Design and conduct a computational experiment to solve an image classification problem (e.g., CIFAR-10 dataset) using a convolutional neural network (CNN) and the AGA technique. First, implement a standard CNN architecture for image classification and train it on the dataset, evaluating its performance on a validation set. Next, apply AGA to automatically generate and optimize CNN architectures. Define a search space of possible network structures, including variations in the number of convolutional layers, filter sizes, and activation functions. Develop an objective function that balances classification accuracy and model complexity. Compare the performance of the AGA-optimized CNN with the standard CNN by evaluating their accuracy and generalization on a test set.

8. Design and experiment to solve an automatic clustering problem, which involves grouping a set of objects so that objects in the same cluster are more similar than those in other clusters, using the AGA technique to maximize a similarity function. Begin by defining the clustering problem as an optimization problem. Then, formulate the master problem considering appropriate penalty functions. Use AGA to automatically generate and optimize clustering algorithms by defining a search space with different clustering approaches and their parameters. Run the AGA process to evolve clustering algorithms over multiple generations, selecting, crossing over, and mutating the best-performing algorithms. Compare the performance of the AGA-optimized clustering algorithms with standard clustering methods by evaluating their similarity scores and computational efficiency.

CHAPTER 6

Producing Metaheuristics Automatically

6.1 METAHEURISTICS AND AGA

The automatic generation of metaheuristics (AGM) consists of automatically designing algorithms with metaheuristic characteristics. This method aims to bypass the reliance on human expertise to devise algorithms tailored to various optimization challenges. Instead of exploring the solution space for a specific problem, AGM conducts a more overarching search across metaheuristics components.

While AGA-generated algorithms might be efficient for their targeted problems, they may not always generalize well. On the other hand, AGM's metaheuristic products retain the capacity to address diverse optimization challenges, given the proper adjustments or modifications. Thus, the AGM and AGA provide distinct outcomes rooted in their design objectives. AGA focuses on producing algorithms tailor-made for specific problem instances or classes. These algorithms might lack the inherent flexibility of metaheuristics, being more specialized and adept at solving particular tasks. Conversely, AGM aims to craft broad algorithmic frameworks suitable for various optimization problems. The resultant structures encapsulate the core attributes of metaheuristics: adaptability, versatility, and general applicability.

DOI: 10.1201/9781003542896-6

AGM aims to identify a metaheuristic that excels across a range of problems or to craft one specifically suited for a particular problem instance. With AGM, it is possible to identify the most promising structure or a combination of them to find an efficient general algorithm capable of solving a complex optimization problem. With AGM, developing more effective and efficient metaheuristics may be possible, particularly for novel or complex problems with limited human expertise. The AGM is a complex task, and many challenges remain, including how to effectively explore the vast space of possible metaheuristics, evaluate and compare metaheuristics, and ensure the generalizability and robustness of the generated algorithms. This chapter shows the steps to generate appropriate metaheuristics efficiently.

A metaheuristic is a computational method that optimizes a problem by iteratively trying to improve a candidate solution regarding a given measure of quality or objective function. While metaheuristics can provide high-quality solutions to many complex optimization problems, there's no guarantee that they will always find the global optimum. They're helpful when the problem is too complex or not well-understood enough to use a more specific algorithmic approach; thus, metaheuristics are not problem-specific. Metaheuristics find optimal or near-optimal solutions for a wide variety of optimization problems, such as scheduling, routing, and assignment problems, as well as machine learning, data mining, and bioinformatics. Generally, a near-optimal solution is also a good solution that balances optimality, feasibility, quality, and efficiency.

Evaluating the effectiveness of a solution in optimization problems requires a multifaceted approach based on several critical criteria. A good solution must meet several criteria. Firstly, it must satisfy all constraints of the optimization problem; any solution that violates these constraints, regardless of its other merits, is generally deemed infeasible. Secondly, the degree to which a solution adheres to the objective function defines its optimality. However, with challenges like NP-hard problems, where finding the exact solution might be computationally challenging, a solution's quality is often gauged by its proximity to the optimal or estimated solution, typically expressed as a percentage deviation. Lastly, efficiency is pivotal, especially in real-time contexts or scenarios with limited computational resources. In such cases, promptly acquiring a reasonably good solution might outweigh the benefits of a marginally superior solution that takes longer to compute.

The necessity of metaheuristics for solving combinatorial optimization problems arises from two main characteristics of these problems. First, many combinatorial optimization problems are NP-hard, meaning no known algorithm can solve all instances of the problem in polynomial time. The number of possible solutions grows exponentially with the size of the problem, making it infeasible to find the optimal solution by brute force. Second, while ad-hoc heuristics may work well for specific instances of a problem, these heuristics often do not generalize well to all instances of the problem. Thus, metaheuristics provide a general framework to guide the search for near-optimal solutions, regardless of the specific details of the problem.

Metaheuristics, as a field of study, has been widely explored and continues to evolve in the research literature. The foundations of metaheuristics were laid in the second half of the 20th century with the development of the first heuristics and metaheuristics, like Genetic Algorithms in the 1960s and Simulated Annealing in the early 1980s. These works established the basic principles of metaheuristic search, such as exploration versus exploitation, intensification versus diversification, and the use of randomness and memory. During the 1980s–1990s, the establishment and consolidation of various fundamental techniques arose, like Tabu Search in the 1980s and Ant Colony Optimization in the 1990s. These techniques introduced new concepts, like the use of long-term memory in Tabu Search and the use of positive feedback in Ant Colony Optimization.

Over the past few decades, the field of metaheuristics has seen significant evolution, characterized by phases of hybridization and theoretical deepening. In the 2000s, as the discipline matured, the emphasis shifted toward blending different metaheuristics and heuristics to craft hybrid algorithms. Notable innovations like the Variable Neighborhood Search amalgamated local search with systematic neighborhood alterations, and applications extended to scheduling, vehicle routing, and machine learning. This hybridization phase was followed by a period from the 2010s to the 2020s marked by a pivot toward theoretical research. Instead of purely empirical studies, there was an intensified push to decipher the behavior and performance theoretically. This era also witnessed the introduction of benchmarking tools and frameworks such as CEC, BBOB for continuous optimization, and PACE for discrete realms, facilitating a standardized comparison of different metaheuristic approaches.

With the exponential advancements in AI over recent years, the spotlight on metaheuristics has intensified. AI's innate ability to process vast datasets and simulate complex scenarios has catalyzed the evolution and application of metaheuristics in solving intricate optimization problems. These heuristic-based strategies, traditionally used to find approximate solutions to complex optimization challenges, are now supercharged by AI's computational prowess. This synergy paves the way for more sophisticated, adaptive, and efficient algorithms. Researchers recognize the potential of integrating AI techniques with metaheuristics, hoping to push the boundaries of what's possible in domains ranging from logistics and supply chain optimization to real-time decision-making in dynamic environments. The convergence of AI and metaheuristics signifies a promising frontier in computational intelligence, opening new avenues for innovation and problem-solving.

It's important to note that while metaheuristics are powerful tools for solving complex optimization problems, they are not always the solution. The "No Free Lunch" theorems have shown that no single optimization algorithm is universally superior. Choosing a suitable metaheuristic depends on the problem, its characteristics, and the available computational resources.

Various metaheuristic approaches have emerged in recent years, each drawing inspiration from diverse natural or artificial phenomena. Each metaheuristic develops independently with distinct principles, methodologies, and research communities. Yet, there lies significant untapped potential in merging the strengths of these individual metaheuristics. Some initial endeavors, like hybrid metaheuristics, have hinted at the power of such combinations, joining approaches like Genetic Algorithms with Tabu Search or integrating Simulated Annealing with Particle Swarm Optimization.

However, merging these distinct methods isn't without its challenges. Each metaheuristic comes with unique parameters, representations, and operators, and their effective integration demands a profound grasp of their nuances. The field could immensely benefit from creating unified frameworks that simplify the fusion of diverse metaheuristics, promoting improved algorithmic performance and a deeper comprehension of shared underlying principles. Beyond mere performance enhancements, this integrative perspective could offer novel insights into the foundational nature of optimization challenges and impulse groundbreaking theoretical progress.

The advent of AGA brings forth a new dimension to the metaheuristic landscape. This technique provides a systematic way to construct and explore an expansive range of metaheuristic combinations, which might have been infeasible or excessively time-consuming when done manually. Instead of relying purely on human intuition and expertise to combine metaheuristics, AGA uses computational power to fuse and tweak different metaheuristic elements systematically.

A salient observation is the potential for elucidating unforeseen synergistic effects. Traditionally, manually engineered hybrid algorithms typically amalgamated well-characterized metaheuristics or components with inherent complementarity. AGA transcends this paradigm, identifying and exploiting synergies between ostensibly disparate metaheuristics. Furthermore, the quantitative assessment of these combinatorial configurations facilitates objective, empirically driven determinations regarding the superior performance of specific hybrid metaheuristics. This approach mitigates the stochastic nature of manual algorithm design, culminating in more productive and potentially innovative solutions. Consequently, AGA leverages computational capacity to revolutionize metaheuristic design methodology, potentially unveiling novel synergies and catalyzing the evolution of optimization techniques.

6.2 TYPES OF METAHEURISTICS

Metaheuristics is a diverse and dynamic field in optimization, offering a range of strategies, each uniquely crafted to address complex solution spaces. These strategies belong to five distinct categories. The first is Single-solution metaheuristics, which focus on continuously refining a single solution. This approach involves an iterative process of enhancement, guided by the problem's objective function and augmented by various heuristics. The second category, population-based metaheuristics, contrasts with the first by working with many solutions, known as a population. These algorithms evolve the population through iterative processes involving selection based on solution fitness, application of variation operators like crossover and mutation, and integration of new solutions.

Further enriching the landscape of metaheuristics are Hybrid Metaheuristics, which combine elements from different methodologies to harness their collective strengths, and nature-inspired metaheuristics, which draw from natural phenomena and behaviors. Finally, the Other Metaheuristics category includes approaches like hyper-heuristics and quantum-inspired algorithms, demonstrating the field's breadth and capacity for continual evolution and innovation. Let us see each category in more detail.

- **Single-solution metaheuristics**. Several distinct traits characterize these techniques. Primarily, they engage in iterative refinement, where algorithms start with an initial solution and methodically enhance it, with adjustments in each cycle often steered by the problem's objective function and supplementary heuristics. They also integrate escape mechanisms to circumvent confinement in local optima by occasionally accepting suboptimal solutions or modifying the search trajectory. Notably, these metaheuristics stand out for their adaptability, aptly for a broad spectrum of optimization problems with slight modifications. Some of them, like Tabu Search, maintain memory structures to record search progress. However, the efficacy of these metaheuristics can be contingent on the problem type, and refining parameters, such as SA's cooling schedule or TS's tabu tenure, are pivotal to achieving optimal outcomes.

- **Population-based metaheuristics.** Those metaheuristics work with a set of solutions, referred to as a population, and evolve them through an iterative process. During each iteration, these algorithms typically select specific solutions based on their quality, apply variation operators to generate new solutions, and then integrate them into the population. Over time, the population's overall quality tends to improve, ideally converging toward an optimal or near-optimal solution to the problem. Examples of population-based metaheuristics include genetic algorithms, particle swarm optimization, and differential evolution.
- **Hybrid metaheuristics.** This family of metaheuristics represents a fascinating and vital frontier in optimization. The inception of hybrid metaheuristics stems from the observation that while a particular metaheuristic might perform exceptionally well for specific optimization problems, it might be suboptimal or inadequate for others. However, when combined with another metaheuristic that offers complementary characteristics, the resulting hybrid can leverage both strengths, offering a more versatile and robust solution approach. There are different ways in which metaheuristics can interact. Some hybrids employ a sequential approach, where one metaheuristic refines the solutions generated by another. For instance, a genetic algorithm might generate diverse solutions, followed by a simulated annealing process to refine and improve them. Alternatively, a more integrated approach can be employed where components from different metaheuristics are more intrinsically intertwined. For example, tabu search strategies might work within a particle swarm optimization framework to enhance exploration capabilities. In this group are metaheuristics such as memetic algorithms that combine a genetic algorithm and a local search; the

greedy randomized adaptive search procedure, combining constructive heuristics with local search.

- **Nature-inspired metaheuristics**. These techniques conceptualize observing and understanding natural processes, phenomena, and behaviors. These metaheuristics leverage the principles that have evolved in nature over millennia, turning them into computational strategies to tackle complex optimization problems. While many metaheuristics take inspiration from nature, these metaheuristics specifically try to mimic natural phenomena, biological processes, or behaviors. Thus, nature showcases a plethora of adaptive and self-organizing systems. From the synchronized movement of bird flocks to the intricate foraging behaviors of ants, these processes have evolved to be efficient and effective. Recognizing the potential of these natural strategies, researchers have abstracted and translated them into algorithmic constructs. To this family belongs the harmony search, which emulates the improvisation of musicians, and the cuckoo search, inspired by the brooding behavior of some cuckoo species. Also, the firefly algorithm takes inspiration from the flashing patterns of fireflies, and the bat algorithm is inspired by the echolocation behavior of bats, among many other exciting metaheuristics.
- **Other metaheuristics**. Given that metaheuristics is a broad and continuously evolving field, there are always methods that might not fit neatly into traditional classifications like single-solution, population-based, hybrid, or nature-inspired. This class encompasses a collection of optimization methods that do not conform strictly to the common metaheuristic categories. They might derive inspiration from non-traditional sources, utilize unique mechanisms, or combine various aspects from different classes in ways that make them hard to classify distinctly. To this class belong the hyper-heuristics that aim to generate or select heuristics for varying problems, making them adaptable across diverse problem domains. Also, the quantum-inspired algorithms draw from quantum mechanics principles.

6.3 KEY CONCEPTS IN METAHEURISTICS

The fundamental components and concepts of metaheuristics serve as the basis for their development and automatically generate new metaheuristics. A thorough comprehension and fine-tuning of these components can substantially impact a metaheuristic algorithm's efficacy, accuracy, and performance when applied to optimization problems. At the core of metaheuristics lies the challenge of navigating the vast and often complex solution landscape to identify the optimal or near-optimal solutions. This challenge is neither arbitrary nor unguided but guided by carefully crafted strategies and principles encoded within the metaheuristic. Consequently, the trajectory from an initial solution to the final is an intricate interplay of various strategies and concepts that aim to traverse the solution space efficiently. The following concepts enable a comprehensive understanding of a given metaheuristic:

- Current Solution: The solution currently being considered or evaluated by the algorithm. Acts as a reference point for exploring new solutions and comparing their quality.
- Neighbor Solution: A solution in the neighborhood of the current solution. It provides alternative solutions for consideration during the search process.
- Neighborhood: For a given solution within the search space, the neighborhood consists of all reachable solutions by making a small, predefined change to the current solution. Exploring the neighborhood is critical for local search methods. It allows the algorithm to move to new potential solutions by adjusting the current solution.
- Acceptance Probability: The likelihood of accepting a solution from the neighborhood, even if it's worse than the current solution. It helps algorithms like simulated annealing escape local optima by occasionally accepting sub-optimal solutions.
- Solution Evaluation: This evaluation typically uses the objective function, which quantitatively measures the solution's effectiveness.
- Initialization: The process of generating a starting solution or set of solutions for the algorithm. It provides a starting point for the search. The initialization method can impact the algorithm's performance, especially for algorithms sensitive to initial conditions.
- Termination Criteria: Conditions determining when the algorithm should stop, such as reaching a maximum number of iterations, a time limit, or a desired solution quality. It provides a clear stopping point for the algorithm and can influence the quality of the final solution obtained.

- Local Search: Local search algorithms iteratively refine a solution by exploring its neighborhood to find a better solution. Helps in iteratively improving solutions, ensuring the progression toward optimal or near-optimal solutions.
- Intensification and Diversification: Focuses the search on areas of the solution space where good solutions appear. Encourages the search to explore different regions of the solution space to avoid local optima. Balancing these strategies ensures a thorough exploration of the solution space while also exploiting known good areas.
- Short-Term and Long-Term Memory: The short-term memory stores recent solutions or changes, often used in strategies like tabu search to prevent revisiting the solutions already visited. Long-term memory retains information about solutions or solution components over a longer duration, informing the search process about historically excellent or bad choices. It helps guide the search and prevent cyclic behaviors, ensuring a more efficient traversal of the solution space.
- Solution Representation: Before a metaheuristic algorithm can work with a solution, it needs to represent it in a way that facilitates manipulation and evaluation. In optimization problems, solution representation might be as simple as a vector of numbers or more complex structures like trees or matrices, depending on the problem domain.
- Genotype-Phenotype Mapping: In evolutionary algorithms inspired by biological processes, the representation of solution occurs in two steps: a genotype that encodes a simplified version of the solution and a phenotype often used to construct the actual solution in terms of the original optimization problem.
- Coding–Decoding: Directly linked to genotype-phenotype mapping, coding refers to transforming a solution (phenotype) into its encoded form (genotype) suitable for manipulation by the algorithm. Decoding is the reverse process: converting the genotype back to a phenotype to assess its quality or implement it.
- Population: In population-based metaheuristics, especially evolutionary algorithms, a group of solutions is maintained and evolved over iterations. This population provides a diverse pool of potential solutions, enabling the algorithm to explore multiple parts of the solution space simultaneously and potentially converge to the global optimum.

- Evolutionary Operators: Certain operations or mechanisms are applied to the genotypes to evolve the solution population and move toward better solutions. They are mainly the selection, crossover, and mutation. Specific solutions from the population based on their fitness are selected. This process ensures better solutions have a higher chance of influencing the next generation. Crossover combines parts of two or more solutions to create new ones. Finally, mutation introduces small random solution changes to maintain diversity in the population and prevent premature convergence to local optima.
- Path Relinking: A strategy that connects two promising solutions by iteratively applying a series of neighborhood moves to find a better solution.
- Multi-Start Strategies: Techniques that start the search process multiple times from different initial solutions, aiming to increase the chances of finding the global optimum.
- Memory-Based Search: Strategies that store and reuse information about previously visited solutions or solution components to guide the search process.
- Ensemble Methods: Techniques that combine the outputs of multiple metaheuristic algorithms or runs to improve the overall performance and robustness of the solution.

6.4 SOLUTION CONTAINER DEFINITION

Following the structure of the master problem for the AGM process, it is necessary to have a solution container as a repository of solutions. This structure is where the terminal actions take place. It provides an organized framework for storing and managing potential solutions, ensuring the algorithm can seamlessly access, evaluate, and modify them. In the broad landscape of metaheuristics, the container stands as a pivotal component, operating as a storehouse for potential solutions. For some methodologies, the container retains the current solution, offering a baseline for comparison as the search iteratively refines its results. In more dynamic approaches, it takes on a more communal role, curating a collection of solutions, serving as a memory bank for past explorations, or even guiding the division and conquering of complex problems by segmenting them.

The true power of the container is its versatility. The container's role is integral, whether maintaining a singular solution, managing a population in a dynamic ecosystem, archiving historical data, or orchestrating segmented problem-solving. Its configuration and operation can directly influence the efficiency and outcome of the search process, making its proper design and understanding essential for anyone delving into the world of metaheuristics.

The solution container helps manage diversity in the search process. Solution containers that can hold a population of solutions are instrumental in preserving diversity. In population-based algorithms, for instance, a container enables the exploration of various regions of the solution landscape. This wide-ranging exploration reduces the risk of premature convergence to local optima and increases the chances of uncovering the global best solution. Furthermore, memory and historical awareness during the search process are crucial. Containers, especially in techniques like Tabu Search, encapsulate this concept by retaining a record of previously explored solutions or solution regions. This recorded history guides the search away from redundant or less fruitful spaces and optimizes the algorithm's trajectory.

The solution container is the dynamic heart of a metaheuristic algorithm. It's a passive storage space and an active arena where creation, evolution, and refinement occur toward optimality. Proper management and understanding of this component shape the efficiency and outcome of the search process. One can closely monitor the convergence of the metaheuristic by observing changes within the container. Vigilantly monitoring the quality of solutions across iterations helps the container assist the algorithm in detecting stagnation or deciding when to terminate the search. This role is crucial for ensuring computational efficiency and preventing endless pursuits in problem spaces.

Another essential concept of metaheuristic is the parallel processing. The structured environment of a container is conducive to parallel evaluations or operations on multiple solutions. For instance, in a parallel metaheuristic approach, several solutions from the container can undergo simultaneous evaluations, accelerating the search.

Example 6.1 Specify a solution container for a solution-based metaheuristic and particularize it to the simulated annealing (SA) metaheuristic.

In a single-solution-based metaheuristic context, the container often simplifies to a simple structure holding the current solution and other relevant attributes or details. For SA, it is necessary to define a container to store the current solution, the best solution found, and the temperature parameter.

a) Current Solution: This is the solution that the algorithm is currently exploring. It could be a vector of numbers, a permutation, a set, or any other structure, depending on the optimization problem at hand.
b) Best Solution: Though SA operates predominantly on the current solution, it can be beneficial to maintain a record of the best solution found thus far to avoid losing it during the search process.
c) Temperature is a critical parameter for SA; as iterations proceed, the temperature decreases, reducing the likelihood of accepting worse solutions. It influences the algorithm's decision-making process.

Example 6.2 Specify a solution container for a population-based metaheuristic and particularize it to a GA.

A population-based metaheuristic requires storing a set of solutions containing the current population. Thus, the solution container's main component consists of collecting potential solutions. In the GA, solutions correspond to chromosomes or individuals. Each individual represents a potential solution to the optimization problem. The size of the population remains constant primarily throughout the algorithm's execution, though the individuals evolve. A given metaheuristic could also consider a population size change during the search process. Thus, the population store structure could be an array or a list. Also, a second structure is required to store a set of solutions selected for reproduction. The new solutions appear after applying crossover and mutation operators. These offspring will replace some or all the individuals in the main structure.

6.5 TERMINALS DEFINITION

The AGM process necessitates functional primitives that operate on solution containers, driving the evolution and adaptability of the search process. These primitives can be categorized as terminals, which directly modify or update the solution set. These are elementary functions that have an immediate impact on the solution container. Their role is critical, as they're responsible for the changes that propel the search forward. Thus, the functions can assemble the terminals to compose a metaheuristic. The functions are non-terminal components that provide the higher-level structure or logic for the metaheuristic. They can nest or embed terminals, and their synergy drives the overall metaheuristic flow.

The machine's task with the AGM is to discover the best combination of ordering functions and terminals. The goal of evolving and assessing different algorithmic structures is to stumble upon a metaheuristic blueprint that excels for a given optimization problem. This dynamic intertwining of terminals and function primitives in various configurations gives rise to diverse and innovative metaheuristic strategies, potentially outperforming traditional hand-crafted approaches.

Generating terminals for specific optimization problems requires a thoughtful strategic approach to ensure the search is effective and efficient. For example, it is necessary to maintain feasibility during the search or to apply penalty terms in the case of unfeasibility. Thus, the following guide helps to design terminals:

- Preservation of feasibility: It's crucial to design terminals that inherently produce feasible solutions or are paired with mechanisms to restore feasibility. For problems where feasibility is complex or challenging, terminals make incremental or local changes with a low likelihood of violating constraints.

- Incorporation of penalty terms: When maintaining feasibility is computationally expensive or challenging, terminals work with penalty functions. These functions penalize infeasible solutions, returning the search to the feasible region. The design of such penalty terms is crucial; they should be stringent enough to deter infeasibility but not so severe that they hinder exploration.

- Exploitation versus Exploration: Terminals should balance the trade-off between exploiting promising areas of the search space and exploring new regions. For instance, some terminals might focus on refining the current best solution (exploitation), while others might introduce more significant changes or randomness to explore different areas (exploration).

- Domain-specific knowledge integration: If domain-specific knowledge or heuristics are available for the optimization problem, it can be beneficial to integrate this into the design of terminals. This knowledge can guide the search in more promising directions or avoid known pitfalls.

- Scalability: Terminals should consider the scalability of the problem. As problem instances grow, the impact of a terminal on the search space might change. Ensuring that terminals remain effective even as the problem size increases is crucial.
- Simplicity and Efficiency: While designing sophisticated terminals is tempting, keeping them as simple and computationally lightweight as possible is often beneficial. Simple terminals are often more interpretable, more accessible to debug, and faster in execution.
- Diversity Generation: Some terminals introduce diversity into the solution container. In population-based methods, diversity is crucial to prevent premature convergence and ensure a broad exploration of the search space.
- Adaptability: Consider designing terminals that can adapt their behavior based on the current state of the search or the observed quality of solutions. For instance, a mutation rate might increase if the search stagnates. It allows its adaptability to the differences encountered while exploring the solution domain.
- Consistency and Predictability: While randomness can benefit exploration, it's also essential to ensure that terminals have a degree of predictability. This predictability helps understand the metaheuristic's behavior and ensures consistent performance across runs. Notes that the presence of random terminals concludes with a random metaheuristic.
- Feedback Mechanism: Incorporate feedback loops that allow the terminals to adjust their operations based on the solutions' recent history or performance. This dynamic adjustment can make the search process more responsive to the nuances of the problem landscape.
- Exact Terminals: Integrating exact methods as terminals in a metaheuristic framework is also possible, especially with time constraints. As Np-hard optimization problems generally cannot be solved quickly with an exact method, an exact method can advance a solution for a limited amount of computer time. In this way, the exact

method could help to find a solution working on the current solution. Thus, the exact method could improve the current solution. By leveraging the inherent problem-solving strategies of exact methods, the search can be guided toward promising regions of the solution space, consequently overcoming local optima that might be challenging for the metaheuristic to navigate.

Example 6.3 Consider the Iterated Local Search Metaheuristic (ILS). Extract from ILS a metaheuristic terminal.

Algorithm 6.1 corresponds to ILS. This high-level metaheuristic leverages the strengths of local search techniques to explore the solution space of combinatorial optimization problems. The function *Local_Search*(s) gives a new solution from the initial solution s. In turn, the function *Perturb*(s) produces a modification of solution s.

Algorithm 6.1 ILS

```
Input:     Maximum number of iterations (n),
   problem instance set; initial solution (s_0)
Output: Best solution (s*)
   s* ← s_0
   s* ← Local_Search(s_0)
   For i = 1 To n Do
         s_aux ← Perturb(s*)
         s_aux ← Local_Search(s_aux)
         If (f(s_aux) < f(s*)) Then s* ← s_aux;
```

The Algorithms 6.2 and 6.3 correspond to two terminals inspired by ILS. The former is a perturbated local search that starts from a specific solution, for example, the current solution in the container of the optimization problem, and iterates until it reaches a predefined stop criterion. The second terminal accomplishes a sequence of local searches to produce a solution. The terminal returns a solution close to the solution found.

Algorithm 6.2 Terminal from ILS

```
Input:     Current solution (s_c); Stop criterion;
Output: New solution (s_n)
   s_aux ← s_c
   Repeat
         s_aux ← Perturb(s_aux)
```

```
        s_aux ← Local_Search(s_aux)
        If (f(s_aux) < f(s_c)) Then s_c ← s_aux
    Until Stop criterion is met
    s_n ← s_c
```

Algorithm 6.3 Terminal from ILS

```
Input:    Current solution (s_c); Stop criterion;
Output: New solution (s_n)
    s_aux ← s_c
    Repeat
        s_aux ← Local_Search(s_aux)
        If (f(s_aux) < f(s_c)) Then s_c ← s_aux
    Until Stop criterion is met
    s_c ← Perturb(s_c)
    s_n ← s_c
```

6.6 DESIGNING TERMINALS FOR AGM

Having identified some key components of metaheuristics, we can propose terminals that, within the AGA framework, enable the generation of new metaheuristics or combinations of them. This approach allows us to harness the best characteristics of each metaheuristic, creating innovative and efficient algorithms. These terminals facilitate a dynamic and adaptive search process by integrating essential operations such as perturbation, local search, acceptance probability, and initialization. The AGA framework's ability to systematically explore and combine these components provides a powerful tool for developing versatile and robust metaheuristic algorithms tailored to various complex problems. The following examples of terminals could help to generate metaheuristics.

- Local search. Algorithm 6.1 describes a local search terminal that Iteratively refines the current solution stored in the solution container by exploring its neighborhood to find a better solution.

Algorithm 6.1 Local search

```
Input: current _ solution
Output: Modified solution
Begin
    improved_solution = perform_local_search
      (current_solution);
     Update solution_container
 End
```

- Tabu Search Neighborhood Terminal: Explores the neighborhood of the current solution and uses a tabu list to prevent cycling back to recently visited solutions. In Algorithm 6.2, the generate_neighborhood function generates a neighborhood of solutions. Replace with actual neighborhood generation logic.

Algorithm 6.2 Tabu terminal

```
Input: current _ solution, tabu _ list, neighborhood _
  size, tabu _ tenure;
Output: best_solution;
Begin
     best_solution = current_solution;
     best_eval = f(current_solution);
     neighborhood = generate_neighborhood(current_
       solution, neighborhood_size);
     For neighbor in neighborhood Do
         IF neighbor NOT in tabu_list THEN
             neighbor_eval = f(neighbor)
             IF neighbor_eval < best_eval THEN
                 best_solution = neighbor
                 best_eval = neighbor_eval
     Update solution_container with best_solution
     current_solution = best_solution
     tabu_list = append(current_solution)
     IF len(tabu_list) > tabu_tenure THEN
         Update tabu_list
End
```

- Perturbation terminal. Algorithm 6.3 describes a perturbation terminal that Introduces a small, random change to the current solution stored in the solution container. For instance, in a combinatorial optimization problem, this perturbation might involve swapping two elements in the solution array, thus enabling the exploration of nearby solution spaces while avoiding local optima.

Algorithm 6.3 Perturbation terminal

```
Input: current _ solution
Output: Modified solution
```

```
Begin
    perturbed_solution = perform_perturbation
      (current_solution)
    Update solution_container
End
```

- Accept with probability terminal. Algorithm 6.4 describes an acceptance with a probability terminal. It determines whether to accept a new solution based on its quality compared to the current solution, incorporating probabilistic acceptance.

Algorithm 6.4 Solution acceptance

```
Input: current _ solution, new _ solution, current _ T;
Output: Modified solution;
Begin
    delta = evaluate(new_solution) - evaluate
      (current_solution);
    prob = exp(delta)/current_T ;
    if delta < 0 or random() < prob
        current_solution=new_solution;
    update solution_container;
End
```

- Mutation terminal. Algorithm 6.5 introduces random changes to the current solution to maintain diversity in the population and avoid premature convergence.

Algorithm 6.5 Mutation terminal

```
Input: current _ solution, mutation _ rate;
Output: Mutated_solution;
Begin
    mutated_solution=perform_mutation
      (current_solution, mutation_rate)
    update solution_container;
End
```

- Crossover terminal. Algorithm 6.6 describes a population-based terminal that combines parts of two or more solutions to create new solutions, promoting the exchange of good traits between solutions. The solution container stores the new solutions according to some replacement policy.

Algorithm 6.6 Crossover terminal

```
 Input: current _ solution, parent _ solutions;
 Output: Offspring solutions;
 Begin
     offspring_solutions =
perform_crossover(parent_solutions)
     update solution_container; # Replacement policy
 End
```

- Crossover SA Terminal: Algorithm 6.7 combines parts of two parent solutions to create offspring and accepts the offspring based on the simulated annealing acceptance probability.

Algorithm 6.7 Adaptive mutation

```
 Input: current _ solution, parent _ solutions,
current _ Temp;
 Output: Offspring solutions;
 Begin
     offspring_solution =
perform_crossover(parent_solutions)
     delta = f(offspring_eval) - f(current_eval)
     if delta < 0 or random() < exp(-delta /
current_Temp)
         current_solutioon = offspring_solution
 End
```

- GRASP Construction Terminal: Constructs a greedy, randomized solution based on a given candidate list and alpha parameter. The terminal constructs a solution using the GRASP approach by creating a restricted candidate list (RCL) and selecting candidates randomly from the RCL. Here, the candidate_list contains the potential candidates for constructing the solution; alpha is a parameter to control the size of the RCL; the solution is the constructed solution using the GRASP approach, and the solution_container stores and updates the current solution.

Algorithm 6.8 GRASP terminal

```
Input: solution _ container, candidate _ list, alpha;
Output: solution_container;
Begin
  WHILE candidate_list DO:
    sorted_candidates = sort(candidate_list)
    limit = max(1, int(alpha *
len(sorted_candidates)))
    restricted_candidate_list = sorted_
      candidates[:limit]
    selected_candidate = random.choice(restricted_
      candidate_list)
    solution = append(selected_candidate)
    candidate_list.remove(selected_candidate)
    Update solution_container;
 End
```

Like these terminals, many others are possible because each metaheuristic has unique logic and mechanisms. For instance, Researchers can create terminals inspired by algorithms like Ant Colony Optimization, Particle Swarm Optimization, Differential Evolution, and more to reflect their distinct strategies for searching and optimization. By incorporating the characteristics of each metaheuristic, these terminals can be assembled within the AGM framework to generate new, innovative metaheuristics. This modular approach allows for the dynamic combination and recombination of various metaheuristic principles, potentially leading to the discovery of highly efficient and robust optimization algorithms.

Solving the master problem to find the best metaheuristic requires data from different optimization problems in the same evolutionary process. For example, a comprehensive study should include ten different NP-hard problems, each represented by multiple problem instances. This extensive numerical experimentation ensures a thorough evaluation of the generated metaheuristics' performance across diverse scenarios. All of these experiments are conducted under the framework of the no-free-lunch theorem, which states that no single optimization algorithm performs best for all possible problems. Hence, the goal is to identify metaheuristics that offer strong performance across various challenging problems, ensuring their general applicability and effectiveness.

6.7 POSSIBLE COMBINATIONS

Consider a new metaheuristic produced by combining ILS and SA. In the space of algorithms, the master problem algorithm used to visit the space of algorithms could find a new algorithm that integrates the cooling mechanism and probabilistic acceptance criteria of SA into the ILS framework. Algorithm 6.4 contains the pseudocode for the combined algorithm.

The metaheuristic begins with an initial solution and applies local search to it. Then, the metaheuristic enters a loop where it iterates a set number of times (n). In each iteration, it perturbs the current best solution and applies local search to this perturbed solution. Instead of accepting the new solution only if it is better, the algorithm now also considers accepting worse solutions. This acceptance is probabilistic, based on the SA acceptance probability. After each iteration, the algorithm reduces the temperature based on the cooling rate, mimicking the cooling mechanism in SA. Then, If the temperature falls below a certain threshold, it is reset to the initial temperature. This reset can help periodically re-introduce the exploration capability of the algorithm.

Algorithm 6.4 Algorithm ILS-SA

```
Input: Maximum number of iterations (n); problem
Instance; initial solution (s0); Initial temperature
(Ti); Final temperature (Tf); cooling _ rate;
Output: Best solution (s*)
Begin
  s*← s0
  s* ← Local_Search(s0)
  T ← Ti
  For i = 1 to n do:
  saux ← Perturb(s*)
  saux ← Local_Search(saux)
  delta ← f(saux) - f(s*)
  If (delta < 0) Then: s* ← saux
 Else:
 acceptance_probability ← e^(-delta / T)
   s* ← saux
  T ← T * cooling_rate
  If T < Tf   Then:
   T ← Ti
  Return s*
End
```

By integrating the explorative features of SA, namely the ability to accept worse solutions and the occasional dynamic cooling schedule, this combined ILS-SA algorithm balances between exploring new regions of the search space and intensively searching around promising solutions. This balance could improve performance over the original ILS and SA algorithms.

6.8 SUMMARY

This chapter studies the AGM, an approach aimed at innovatively designing algorithms that embody metaheuristic characteristics without extensive reliance on human expertise. AGM seeks to autonomously navigate the vast array of metaheuristic components, identifying and synthesizing those that promise enhanced performance across diverse optimization challenges. Unlike AGA, which produces algorithms specifically tuned to particular problems, AGM aspires to develop versatile algorithmic frameworks that can adapt to various optimization problems. The exploration into AGM highlights its potential to craft algorithms with broad applicability and underscores metaheuristics' intrinsic adaptability and general applicability. This chapter thoroughly examines the foundational aspects of metaheuristics, including crucial concepts like solution representation, neighborhood exploration, and the critical roles of solution containers and terminals in the metaheuristic design process.

The exploration of AGM in this chapter underscores the significant potential of automatically generating metaheuristics helping to solve complex optimization problems. AGM aims to pinpoint effective structures or combinations across various problem domains by systematically exploring and combining metaheuristic components. This approach enhances the efficiency and effectiveness of metaheuristics, particularly for novel or intricate challenges where human expertise may fall short, and illuminates the path toward discovering new algorithmic synergies. Furthermore, AGM presents a promising avenue to streamline and potentially accelerate research within the field of metaheuristics. The periodic emergence of new metaheuristics, which often bear algorithmic similarities to existing ones, underscores a broader challenge: the effort to find innovative solutions can sometimes result in a proliferation of approaches with marginal differences or improvements.

EXERCISES

1. Formulate the master problem to automatically generate a novel metaheuristic to solve the traveling salesman problem. Define the search space, objective function, and constraints. Describe how the genetic programming framework will explore this space to identify effective metaheuristic structures.

2. Define a solution container for a population-based metaheuristic applied to the Vehicle Routing Problem. Specify how the container manages multiple solutions, stores historical data, and supports parallel evaluations to enhance the search process.

3. Design a set of terminals and functions for a genetic algorithm to solve the Binary Knapsack Master Problem. Explain how these primitives will interact within the algorithm's structure to balance exploration and exploitation effectively. Provide an example syntax tree.

4 Develop a hybrid metaheuristic that combines simulated annealing (SA) elements and particle swarm optimization (PSO) to solve the job shop scheduling problem. Model the interaction between SA's temperature-based exploration and PSO's velocity updates. Define the initial setup and integration strategy.

5. Model intensification and diversification strategies for a genetic algorithm (GA) aimed at solving the quadratic assignment problem (QAP). Explain how these strategies can be encapsulated and incorporated as elementary components to solve the master problem.

6. Construct an adaptive metaheuristic for the bin packing problem incorporating automatic parameter tuning within the master problem framework. Define the feedback mechanism that guides parameter adaptation to improve solution quality and convergence speed. Model the search space for the metaheuristic parameters, develop an objective function to evaluate the performance of different parameter settings, and describe how genetic programming will explore this space to identify optimal parameter configurations.

7. Design an experiment to evaluate the performance of an automatically generated metaheuristic for the graph coloring problem. Define the evaluation criteria, including solution quality, computational efficiency, and robustness. Explain how to benchmark the generated metaheuristic against traditional approaches.

8. Propose a method that combines components from metaheuristics simulated annealing, tabu search, and iterated local search to solve the maximum independent set problem. Define the roles of each component, the interaction between them, and the expected benefits of this combination. Model the combined algorithm and its application to the problem.

CHAPTER 7

AGA with Reinforcement Learning

7.1 INTRODUCTION

This chapter delves into the intersection of dynamic programming (DP) and reinforcement learning (RL), two pivotal areas of computational intelligence that allow us to advance the automatic generation of algorithms. The foundational work by Richard Bellman in the 1950s established a mathematical framework integral to modern RL algorithms. This chapter aims to unpack these complex topics, illustrating how DP principles are related to the development of RL and their application to the automatic generation of algorithms. DP is explored through its mathematical and computational facets, emphasizing the breakdown of complex decision-making processes into manageable subproblems. This method has proven effective across various optimization domains. We explore Bellman's principle of optimality and its application in crafting efficient algorithms that scale from theoretical constructs to practical applications.

In parallel, we discuss how RL has evolved from these foundational concepts into a robust framework capable of addressing complex and dynamic environments. By integrating advances in machine learning and computational capabilities, RL has expanded its reach, demonstrating significant capacity in solving intricate optimization problems through learned experiences and interactions. The chapter will mainly focus on modeling the automatic generation of algorithms as a deciding agent problem (DAP), showcasing how DP and RL can be instrumental in

DOI: 10.1201/9781003542896-7

devising algorithms that adapt and excel in a wide array of settings. The DAP is a fundamental problem for modeling practical dynamic situations. Fundamental problems are commonly used in combinatorial optimization to model practical situations. Through this lens, we aim to provide a comprehensive overview that educates and empowers readers to harness these techniques in their endeavors.

To ground these concepts in a practical example, we illustrate how to apply RL to the AGA master problem through the binary knapsack problem. This example demonstrates a technique that can be used in any combinatorial optimization problem, showcasing the versatility and applicability of RL in solving complex optimization challenges. By modeling the problem as a DAP, we show how RL can adapt and excel in various settings, providing a robust method for algorithm generation.

7.2 DYNAMIC PROGRAMMING

Dynamics programming is a mathematical and computational technique for solving complex problems by breaking them into more straightforward, manageable subproblems. Each subproblem is a smaller instance of the original problem, often with a reduced scope or size. Solving one subproblem involves solving several smaller subproblems. A directed graph enables us to represent the solution process as a path consisting of states (nodes) and decisions (arcs).

- States. A state represents a specific condition or situation in the context of a problem. The state encapsulates all the information necessary to decide at that point and solve the subproblem. It often includes the current position, resources, or other relevant attributes.
- Decisions. A decision is an action or choice made in a particular state. Decisions are made at each state to transition to the next state. The choice of decision at each state affects the outcome and the path to the solution. DP aims to find the sequence of decisions that leads to the optimal solution.

To guide the search for the solution, a strategy or an objective function is required that dictates the decisions to adopt in each state. In the context of DP, a policy defines which decision should be made in each state to achieve the best outcome. An optimal policy is one that, when followed, results in the best possible outcome (e.g., maximum profit, shortest path, minimum time) from the given problem. The optimal solution to the problem results from piecing together the optimal solutions of these subproblems.

Approaching an optimization problem using DP requires constructing a model that accurately represents the problem. Similarly, in most optimization problems, this model is a framework for a DP algorithm to find the optimal solution. The components of a DP model are crucial for defining the problem's structure and the rules that govern the solution process. In this way, the model construction requires the definitions of the Decision Space, State Transition Function, Objective Function, and Cost Structure. These model components are as follows:

- State: A state encapsulates all relevant information needed to decide at that point in the problem and often represents a snapshot of the problem at a particular stage, including all necessary information to proceed with decision-making. The state space defines all possible states in which the system can exist.

- Decision Space. It comprises all possible actions or decisions in each state. Decisions made at each state lead to transitions to other states. Thus, the decision space defines the set of all possible actions or decisions in each state. The decision space defines the "what" (what choices are available).

- State Transition Function. Describes how the system transitions from one state to another based on the decisions made. Provides the mechanics of how these choices affect the state of the system. Thus, this function is about the "how" (how these choices lead to new states).

- Cost Structure. Specifies the immediate cost or reward of making a particular decision in each state. In a minimization problem, it could represent the cost incurred, while in a maximization problem, it could represent the reward gained.

- Objective Function. It is a component that quantifies the optimization goal, such as minimizing cost or maximizing profit. It evaluates the quality of the decisions made and is used to determine the optimality of a policy. This broader measure evaluates the overall performance of a policy across all states and decisions; thus, this is the aggregation of the individual costs or rewards aligned with the goal of the optimization. The cost structure feeds into the objective function. Each decision's immediate cost contributes to the cumulative measure the objective function seeks to optimize.

- Boundary Conditions. These are the conditions that define the start and end points of the problem. They include initial conditions and any terminal conditions that signify the completion of the process.

The decision space and state transition function are about the operational aspects of the problem, while the cost structure and objective function are about the evaluative aspects. Thus, the decision space and state transition function work together to define the dynamics of the problem, pointing out what decisions are possible and how they change the system's state. On the other hand, the cost structure and objective function define the optimization aspect, i.e., the consequences of decisions and how they aggregate to meet the overall goal. A feasible solution to the problem is a policy or strategy that specifies the decision to be made in each state, and mainly, the optimal solution or optimal policy is the one that optimizes the objective function.

To comprehensively understand the DP process, drawing an analogy with the shortest path problem, a fundamental and classical optimization issue in graph theory, is helpful. The shortest path problem entails identifying the least-cost route from an initial point to a destination within a network of interconnected nodes. Each link between nodes carries an associated cost, such as distance or time. The objective is to find the path that minimizes the total cost of traveling from the start node to the target node. This problem is prevalent in various fields, including transportation, logistics, and network routing. In the context of a shortest path problem, the DP process consists of finding the optimal path from a starting node (initial state) to a target node (goal state) by making a series of decisions (choosing edges) that minimize the total cost (e.g., distance, time). Thus, each state in this problem corresponds to a specific node in the graph, encapsulating the information about the current position. The decisions at each state involve selecting the next node to move to based on the available edges. The state transition function describes how moving along an edge (decision) leads to a new state (next node).

The cost structure specifies the immediate cost associated with each edge, and the objective function aims to minimize the total cost from the start node to the goal node. Boundary conditions define the start node (initial condition) and the goal node (terminal condition). DP systematically evaluates all possible paths by solving subproblems, such as finding the shortest path to each intermediate node and combining these solutions to determine the shortest path. This procedure involves constructing a DP table where each entry represents the minimum cost to reach a node, updating the table iteratively based on the state transition function and cost structure, and finally extracting the optimal path that achieves the objective of minimizing the total cost from the start node to the goal node.

To model an optimization problem via DP, one must clearly define the states, decisions, how states transition based on these decisions, the objective to be optimized, the immediate costs or rewards of decisions, and the boundary conditions of the problem. The DP algorithm then systematically explores the feasible space of solutions, solving subproblems and combining their solutions to find an optimal policy that achieves the best overall outcome according to the objective function.

7.3 BELLMAN'S PRINCIPLE OF OPTIMALITY

The principle of optimality in DP is a fundamental concept in this field. Introduced by Richard Bellman, this principle states that an optimal policy has the property that, regardless of the initial state and initial decision, the remaining decisions must constitute an optimal policy related to the state resulting from the first decision. Thus, this principle asserts that if you have found the optimal path or strategy to solve a problem, then any sub-path (a portion of the path) of this optimal path is also optimal for the sub-problem that starts at the beginning of this sub-path. For example, consider a problem where you need to travel from point A to point B, then to point C, and finally to point D, and you want to find the shortest path. According to the principle of optimality, if the path from A to D via B and C is the shortest path from A to D, then the path from B to D via C must be the shortest path from B to D. In the context of DP, this principle breaks down a complex problem into simpler sub-problems. The optimal solution to the overall problem is constructed by solving the sub-problems optimally and integrating their solutions.

Consider Figure 7.1, which describes the two graphs that capture the essence of DP and optimization problems in a visual format. The first graph represents the solution space of the shortest-path problem. Each node represents a specific location, such as a city or a checkpoint, and the arcs connecting the nodes represent the paths between locations.

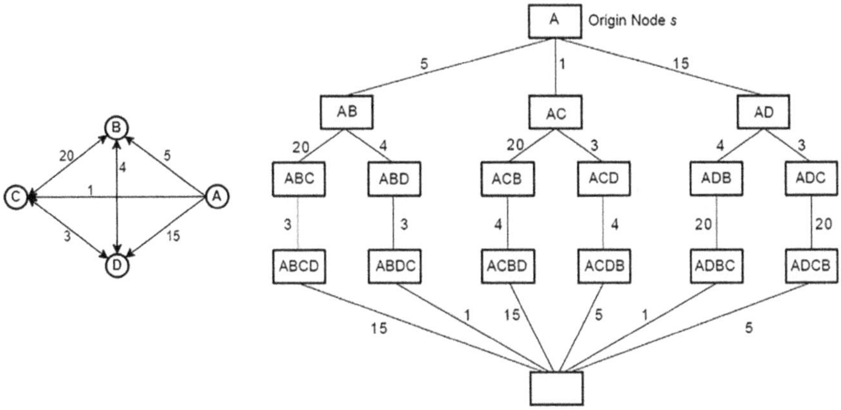

FIGURE 7.1 A graph view of a dynamic programming problem.

The numbers on the arcs represent the cost or distance of traveling between two nodes. The goal is to find the shortest path from the starting node to the destination node, minimizing the total cost or distance. The second graph represents a problem state space. Each node represents a state or stage in the solution process of an optimization problem, and the arcs reflect a decision or transition from one state to another, with associated costs or profits. A specific node represents a current or final state in the problem-solving process. The aim is to navigate through this state space to reach the final node while optimizing a cost function, which could mean maximizing the profit or minimizing the costs, depending on the problem's context. A path through this graph from the initial state to the final state represents a sequence of decisions or transitions, the optimality of which depends on the cumulative cost or profit.

The shortest path graph is a model to visualize the whole solution space, capturing all feasible solutions and allowing for the identification of the optimal one by exploring the paths through the graph. DP solves a similar problem by breaking it into subproblems corresponding to the graph's nodes. The edges connecting the nodes represent the recomposition step, where the solutions to subproblems are combined according to the principle of optimality to form the solution to the main problem. The optimal solution to the main problem is obtained from the optimal solutions to these subproblems.

Formally, consider a decision problem where you must make a sequence of decisions over time to optimize some objective, either minimizing cost or maximizing profit. Let's denote the system's state at time t as s_t and a decision made at time t as a_t. The objective is to find a sequence of decisions $a_1,a_2,\ldots,a_T$, that optimizes the cumulative objective function over a time horizon T. Bellman's Principle of Optimality states that if a decision sequence $\{a_1^*, a_2^*, \ldots, a_T^*\}$ is optimal, then, for any intermediate stage t, the subsequence $\{a_t^*, a_{t+1}^*, \ldots, a_T^*\}$ must be an optimal decision sequence for the problem starting from the state s_t following the decisions $\{a_1^*, a_2^*, \ldots, a_{t-1}^*\}$.

If we let $V(s_t)$ be the value function representing the optimal cumulative objective for a minimization problem from state s_t to the end of the horizon and $T(s_t,a_t)$ be the transition function that gives the next state given the current state s_t and decision a_t, then, Bellman's equation is given in Equation (7.1). This equation essentially states that the value of being in a particular state s_t is the cost of the best decision a_t at that state, plus the value of the state that this decision leads to.

$$V(s_t)=\min_{a_t}\left[C(s_t,a_t)+V\left(T(s_t,a_t)\right)\right] \tag{7.1}$$

This formulation highlights that the optimal value $V(s_t)$ is derived by minimizing the sum of the immediate cost $C(s_t,a_t)$ and the optimal value of the subsequent state $V(T(s_t,a_t))$.

7.4 DYNAMIC PROGRAMMING ALGORITHMS

Bellman's equations provide the recursive relationship that underpins the DP approach. In the context of a graph that represents the solution space, a DP algorithm acts as a traversal technique. Bellman's equations allow us to navigate through a graph containing all possible solution paths, solving subproblems optimally and using those solutions to address progressively larger subproblems of the main problem until the overall problem is solved. This approach ensures that the traversal occurs in a way that guarantees the most efficient use of computational resources, typically turning an intractable problem into a tractable one. Thus, a DP algorithm traverses this graph not by exploring every possible path exhaustively but by systematically solving subproblems (nodes) and using their solutions (the optimal paths leading to them) to build up the solution to larger subproblems. The Bellman's equation dictates the optimal structure of the traversal. As a DP algorithm traverses the graph, it stores the solutions to subproblems so that each subproblem is solved only once. This process is crucial for ensuring that DP is efficient, avoiding the exponential time complexity resulting from naive recursive solutions. Once the traversal is complete, the algorithm can reconstruct the optimal solution or policy by backtracking from the terminal state (the goal node) through the optimal predecessors, defined by the solutions stored during the traversal. Throughout this process, a DP algorithm adheres to the principle of optimality, which states that the optimal solution to any subproblem can be composed of the optimal solutions to its subcomponents.

A DP algorithm traverses a solution space graph by decomposing the overarching problem into overlapping subproblems, solving each at once, and caching their solutions in a structured table to prevent redundant computations. This process ensures that subsequent encounters of a subproblem leverage the cached solution, enhancing computational efficiency. In the context of a shortest-path graph traversal, for instance, DP exploits this principle by iteratively building up the minimum travel cost from the origin to all nodes, utilizing previously computed costs stored to determine the optimal path to any intermediary node, thereby constructing the global optimum from these local optima.

Each DP algorithm employs one of two search mechanisms: bottom-up and top-down. The bottom-up approach, also known as iterative DP, starts by first solving the smallest subproblems and gradually building up solutions to larger subproblems using previously computed results until the final solution is reached. This method typically uses a table to store the results of subproblems, ensuring that each subproblem is solved only once, thus avoiding redundant computations. On the other hand, the top-down approach, often called memoization, begins by trying to solve the largest problem and recursively breaks it down into smaller subproblems, storing the results of these subproblems to avoid redundant calculations in future recursive calls. This method combines the advantages of recursion with the efficiency of storing intermediate results, enhancing the algorithm's overall performance. Both approaches ultimately aim to optimize the problem-solving process by efficiently managing the computation of subproblems and their contributions to the overall solution. Algorithm 7.1 represents a top-down induction approach that starts from the last stage and moves backward to the first stage, updating the value function and policy at each step. The cost and transition functions represent the immediate cost and the state transition dynamics, respectively. This general algorithm can be written for a maximization problem, replacing the word cost with the word reward.

Example 7.1

Let's use a simple example to illustrate how the General DP Algorithm works. We will solve a DP problem as a small, directed graph with four states (A, B, C, D) and a time horizon of 3 stages. The objective is to find the best policy from node A to node D. Node A is the initial state, and node D is the goal state. The transition costs are A to B: cost = 1; A to C: cost = 4; B to C: cost = 2; B to D: cost = 6; and C to D: cost = 3. Figure 7.1 depicts the graph and the ValueFunction table for the algorithm execution.

Algorithm 7.1 General Dynamic Programming Algorithm

Input :

States: A set of all possible states;
Actions(state) : Function returning a set of possible actions for a *state*;
Transition(state, action) : Function returning the next *state* given a current *state* and *action*;
Cost(state, action) : Function returning the immediate *cost* of taking an *action* in a *state*;
Horizon: The number of stages or time steps in the problem;

Output :

ValueFunction: A Table containing the minimum cost achievable from each state;
Policy: A Table indicating the best *action* to take in each *state*;

Begin

Initialize *ValueFunction* for all states, typically to a large number or some initial guess;
Initialize Policy for all states to a null action;

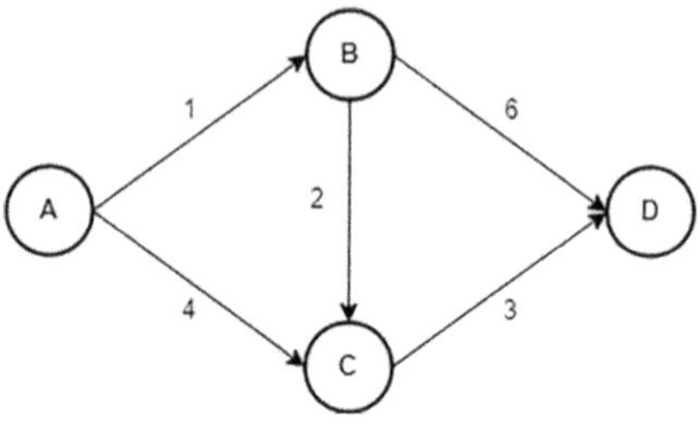

Node	Init	Iter 1, stg3	Iter 2, stg 2	Iter 3, stg 1
A	inf/null	inf/null	5/AB	5/AB
B	inf/null	6/BD	5/BC	5/BC
C	inf/null	3/CD	3/CD	3/CD
D	inf/null	0/null	0/null	0/null

FIGURE 7.2 Representation of the iterative process for Example 7.1.

```
For each stage from Horizon down to 1 Do:
    For each state in States Do:
         BestValue = M (A large number for a
         minimization problem);
         BestAction = null;
       For each action in Actions(state) Do:
            NextState = Transition(state, action)
            Value = Cost(state, action) +
            ValueFunction[NextState]
         If Value < BestValue Then
            BestValue = Value
            BestAction=action
            Update Policy for the current state of
                  this action
         ValueFunction[state] = BestValue
         Policy[state] = BestAction
  Return ValueFunction and Policy
End.
```

7.5 DYNAMIC PROGRAMMING APPROACHES

DP considers deterministic and stochastic approaches. Deterministic DP works in environments with specific and predictable outcomes, such as resource allocation or shortest-path problems. The state transition is known in these scenarios and does not involve randomness, making the problem-solving process more straightforward. Stochastic DP deals with environments where outcomes involve uncertainty and probabilistic state transitions. This approach is essential in scenarios like inventory management with uncertain demand or investment planning under market volatility. Stochastic DP requires calculating expected values and probabilities, making it more complex than its deterministic counterpart.

DP is also categorized based on the time horizon of the problem: finite or infinite. Finite horizon problems have a known and fixed number of stages, such as in multistage investment decisions or project planning with a definite end date. Infinite horizon problems, in contrast, extend indefinitely and are common in long-term planning like financial management or resource allocation. These problems often focus on finding a steady-state or long-term optimal policy, using methods like value iteration or policy iteration to find convergent solutions. Thus, the combinations give rise to four categories: a) Deterministic DP—Finite horizon problems, b) Deterministic DP—Infinite horizon problems, c) Stochastic DP—Finite horizon problems, and d) Stochastic DP—Infinite horizon problems. Each category presents unique challenges and requires specific methodologies for finding optimal solutions.

The deterministic cases are generally more straightforward, as they do not involve uncertainty. In contrast, stochastic cases require dealing with probabilities and expected values, adding layers of complexity. The time horizon also significantly influences the problem-solving approach, with finite horizon problems often using backward induction and infinite horizon problems focusing on finding steady-state or convergent solutions. Specifically,

a) **Deterministic and Finite Horizon Problems (DFHDP).** These problems involve a known and fixed number of stages, and the outcomes of all decisions are specific and predictable. Examples include planning problems where each step has a definite outcome and a known endpoint, like scheduling tasks over a fixed period.

Example 7.2

A manufacturing company must plan its production schedule for the next four months. The goal is to meet the forecasted demand for its product while minimizing the total cost, which includes production costs, inventory holding costs, and penalties for not meeting demand. The demand, production capacity, and costs are known and do not change unpredictably.

Knowing that the period is four months and the demand is perfectly predictable, the problem can be approached as a DFHDP problem. Let's consider the following notation:

D_t: Demand in month t.
P_t: Production in month t less than or equal to the production capacity, $P_t \geq 0$.
I_t: Inventory at the end of month t.
$C(P_t)$: Production cost for producing P_t units.
H (It): Holding cost for having It units in inventory at the end of the month.
$F(I_t, D_t)$: Penalty cost for failing to meet demand.
Cap: Production capacity.

Formulation:

- Equation (7.2) corresponds to the inventory balance, and Equation (7.3) represents the fulfillment of each variable's limit values.

$$I_{t-1} + P_t - D_t = I_t \forall t = 1,2,3,4 \tag{7.2}$$

$$\text{Cap} \geq P_t \geq 0 \,\text{and}\, I_t \geq 0 \forall t = 1,2,3,4. \tag{7.3}$$

- State: The state at the beginning of month t is given by the inventory level I_{t-1}.
- Decision Variable: The decision variable at each stage t is the production quantity P_t.
- Cost Function: The cost at each stage t is given by:

$$C(P_t) + H(I_t) + F(I_t, D_t) \tag{7.4}$$

- Thus, the PD problem is:

$$J_3(I_3) = \text{Min}\left[C(P_4) + H(I_4) + F(I_4, D_4)\right] \tag{7.5}$$

$$J_t(I_{t-1}) = \text{Min}\left[C(P_t) + H(I_t) + F(I_t, D_t) + J_{t+1}(I_t)\right] \text{for}\, t = 1,2,3. \tag{7.6}$$

Subject to:

$$I_{t-1} + P_t - D_t = I_t; \text{Cap} \geq P_t \geq 0 \text{ and } I_t \geq 0, \forall t = 1,2,3,4. \tag{7.7}$$

An algorithm for a deterministic DP problem with a finite horizon typically involves backward induction, starting from the last stage and moving backward to the first stage, calculating the optimal decision at each stage.

b) **Deterministic DP—Infinite Horizon Problems.**
The problems extend indefinitely in this case, but the outcomes remain predictable. An example could be a long-term investment strategy where the returns are deterministic, and the planning horizon is infinite.

Example 7.3

A company decides whether to continue operating or replace existing equipment with new equipment annually. The goal is to minimize long-term costs, including maintenance costs for existing equipment and purchasing new equipment. The costs are known and deterministic. The problem extends indefinitely (infinite horizon), as the company will face this decision every year. Let's consider the following notation:

C_t: Cost of operating the equipment in year t, which increases with the age of the equipment.
R: Cost of purchasing new equipment.
A_t: Age of the equipment in year t.

Formulation:

- State: The state in year t is given by the age of the equipment A_t.
- Decision Variable: The decision variable at each stage t is whether to replace the equipment (binary decision: replace or not).
- Cost Function: The immediate cost in year t is $C_t(A_t)$ if the equipment is not replaced, or R if it is replaced.
- Recurrence Relation: The cost-to-go function $J(A_t)$ represents the total cost from year t onwards, given the age A_t of the equipment.

$$J(A_t) = \min C_t(A_t) + J(A_t + 1), R + J(0) \tag{7.8}$$

This equation states that the total cost is the minimum cost of operating the equipment for one more year and then following the optimal policy or the cost of replacing the equipment and then following the optimal policy.

The problem can be solved using value iteration or policy iteration methods. These methods iteratively update the cost-to-go function $J(A_t)$ or the policy until it converges to a steady-state solution. Since the problem has an infinite horizon, the focus is on finding an optimal stationary policy regardless of the starting state.

c) **Stochastic DP—Finite Horizon Problems.**
These problems have fixed stages, but the outcomes are uncertain and involve probabilistic state transitions. In a multistage investment decision under market uncertainty, investors face uncertain outcomes at each stage but must adhere to a fixed overall investment horizon.

Example 7.4

A retailer must decide how much inventory to order each month for the next six months (finite horizon). The demand for the product is uncertain and varies each month (stochastic). The goal is to minimize the total cost, which includes ordering costs, holding costs for excess inventory, and shortage costs for unmet demand. Propose a model representing this situation as a Stochastic DP problem with finite horizon. Let's consider the following notation:

D_t: Random demand in month t.
Q_t: Order quantity in month t.
I_t : Inventory level at the end of month t.
$C(Q_t)$: Cost of ordering Q_t units.
$H(I_t)$: Holding cost for inventory I_t.
$S(I_t, D_t)$: Shortage costs if demand exceeds inventory.

Formulation:

- State: The state at the beginning of t is the inventory level I_{t-1}.
- Decision Variable: The decision variable at each stage t is the order quantity Q_t.
- Cost Function: The immediate cost in month t is $C(Q_t) + H(I_t) + S(I_t, D_t)$, where $S(I_t, D_t)$ depends on the random demand D_t .
- Recurrence Relation. The expected value gives the cost-to-go function:

$$J\{t\}(I_{t-1}) = E\{C(Q_t) + H(I_t) + S(I_t, D_t) + J\{t+1\}(I_t)\} \forall t = 1,2,\ldots,5 \tag{7.9}$$

subject to:

$$J\{6\}(I_5) = E\{C(Q_6) + H(I_6) + S(I_6, D_6)\} \tag{7.10}$$

$$I_{t-1} + Q_t - D_t = I_t, \forall t = 1,2,\ldots,6. \tag{7.11}$$

$$Q_t \geq 0,\, I_t \geq 0 \;\forall t = 1,2,\ldots,6 \tag{7.12}$$

The problem can be solved using backward induction, starting from the last month (month 6) and moving backward to the first month. At each stage, the procedure calculates the optimal order quantity to minimize the expected total cost, considering the stochastic nature of demand. This procedure involves calculating the expected costs for different order quantities at each stage, taking into account the probability distribution of the demand D_t.

d) **Stochastic DP—Infinite Horizon Problems**
These problems have no fixed endpoint, and the outcomes involve uncertainty. A typical example is long-term financial planning or resource management under conditions of uncertainty, where the goal is to find an optimal policy that remains effective indefinitely.

Example 7.5

A company owns equipment that requires ongoing maintenance and occasional replacement. The equipment's failure probability increases with age and usage, making future performance uncertain. At the end of each year, the company must decide whether to maintain the equipment, replace it with new equipment, or do nothing. The goal is to minimize the long-term expected cost that can be considered a problem with an infinite horizon. This problem aims to minimize the expected long-term cost of equipment maintenance, replacement, and failure. Thus, decisions must be made annually based on the equipment's age and condition, and the age will reset to zero if it is replaced. Let's consider the following notation:

A_t: Age of the equipment in year t.
X_t: Decision in year t (maintain, replace, or do nothing).
$C_m(A_t)$: Maintenance cost for the equipment at age A_t.
C_r: Cost of purchasing new equipment.
$F(A_t)$: Failure cost if the equipment fails, which depends on its age.
$J(A_t)$: The expected total cost from year t onwards, given the age A_t of the equipment.

Formulation:

- State: The state in year t is the age of the equipment A_t .
- Decision Variable: the action X_t at each stage t (maintain, replace, or do nothing).
- Cost Function: The immediate cost in year t depends on the decision X_t and the age A_t. It includes maintenance cost, replacement cost, or failure cost.

- The recurrence relation states that the total cost is the minimum of the cost of maintaining the equipment for one more year, replacing the equipment, or doing nothing and facing a potential failure.

$$J(A_t) = \text{Min}\, E\left[C_m(A_t) + J(A_{t+1}), C_r + J(0), F(A_t) + J(A_{t+1})\right] \quad (7.13)$$

7.6 DECIDING AGENT PROBLEM

A repertoire of fundamental problems, such as those encountered in combinatorial optimization, is beneficial when modeling complex real-world issues using DP. These canonical problems provide a scaffold for developing algorithmic strategies that can either be applied directly or with modifications to suit specific real-world scenarios. When a fundamental problem aligns closely with a practical situation, it allows for a straightforward application of existing DP models, which can be calibrated and validated relatively quickly. However, real-world problems often exhibit nuances and intricacies that these fundamental problems do not capture fully. In such cases, the foundational problem is an approximation or a simplified version of reality. Including additional layers of complexity into the model, one can bridge such a gap that may involve augmenting the state space, refining the action set, or adjusting the reward structure and transition dynamics. Also, it could be necessary to extend the model to account for uncertainty, multiagent interactions, or adapting to temporal variations.

We define a theoretical optimization problem called the Deciding Agent Problem, in which an abstract agent dynamically makes decisions to maximize the whole reward obtained during a period. Specifically, given an environment with defined state and action spaces, transition dynamics, and a reward function, the goal is to find an optimal policy for the agent that maximizes (or minimizes) the objective function over a series of decisions. The agent must effectively balance the search process to improve its decision-making strategy over time, especially in the face of environmental uncertainty and dynamics. The DAP can be stated as follows:

Given:

a) A finite set of states S representing all possible configurations of the environment.

b) A finite set of actions A, representing all possible decisions the agent can make.

c) A transition dynamics $\boldsymbol{T}: \boldsymbol{S} \times \boldsymbol{A} \times \boldsymbol{S} \rightarrow [\mathbf{0},\mathbf{1}]$ a function that defines the probability $T(s,a,s')$ of transitioning from state s to state s' given action a. This encapsulates the stochastic nature of the environment.

d) The transition probabilities must satisfy $\sum_{s' \in S} T(s,a,s') = 1$ for each $s \in S$ and $a \in A$.

e) A reward function $\boldsymbol{R}: \boldsymbol{S} \times \boldsymbol{A} \times \boldsymbol{S} \rightarrow \mathbb{R}$, which assigns a real-valued immediate reward (or cost) for transitioning from state s to state s' due to action a.

f) A discount factor $\gamma \in [\mathbf{0},\mathbf{1}]$, which models the preference for immediate rewards over future rewards.

Then:

The objective is to maximize the expected return G from the initial state s_0, where G is the sum of discounted rewards over time, and as a result, the policy $\pi{:}S{\rightarrow}A$ that maximizes the expected cumulative discounted reward over time is found.

$$G = \sum_{t=0}^{\infty} \gamma^t R(s_t, \{\pi(s\}_t), s_{t+1}) \tag{7.14}$$

This formal statement captures the essence of the DAP in a mathematical framework, highlighting the key components of the state, action, transition dynamics, reward function, and the objective of maximizing the expected cumulative reward.

The DAP serves as a versatile framework for modeling practical situations such as DP problems, particularly when the structure of the problem aligns with phenomena that change over time. The essence of a DAP lies in finding a policy that either maximizes or minimizes a cumulative metric, such as reward or cost, which is a common objective in DP. When a DAP has a finite and well-defined state and action space, DP can effectively address the problem by systematically evaluating the outcomes of various actions across different states to determine the optimal policy. With known transition dynamics and reward functions, DP can leverage this information to compute the value of different states and actions, which is crucial for policy determination. The Bellman equations recursively decompose the value function, a technique that fits well with the structure of many DAPs. This recursive decomposition allows for the expression of future values as a function of the subsequent state's value and the immediate reward. Moreover, DP is applicable even in scenarios involving uncertainty or stochastic elements, making it a robust tool for a wide range of real-world DAPs where outcomes are not deterministic. We can use DP to systematically explore and optimize decision-making strategies by framing practical problems within the DAP framework.

The DAP, like the binary knapsack, traveling salesman, or graph coloring problems, represents a fundamental class of problems in decision-making and optimization. We consider these problems "fundamental" because they capture the essence of a wide range of practical challenges across various domains. The DAP encapsulates the core challenge of making a sequence of decisions to optimize a specific objective in a dynamic and uncertain environment, a scenario common in many real-world applications. Furthermore, different fundamental problems often require different analytical frameworks for their solution. For instance, linear and integer programming frameworks are well-suited for optimization problems with linear relationships and constraints. DP is effective in scenarios involving sequential decision processes with optimal substructure, decomposing these problems into overlapping subproblems. Each of these frameworks includes a range of algorithms designed to solve problem instances precisely or approximately, providing flexibility and adaptability in addressing various complexities and requirements.

Many practical problems correspond to instances of these fundamental problems. Modeling is critical because it translates real-world challenges into a formal structure that established analytical methods can address. Once we appropriately model a practical problem, we can apply suitable algorithms from the corresponding framework to find exact or approximate solutions. While we can solve some problems precisely using these frameworks, others may be too complex or computationally intensive for an exact solution. We turn to approximation, metaheuristics, or automatically generated algorithms in such cases. These methods aim to find good enough solutions within reasonable computational time, often crucial in practical applications.

7.7 AGA WITH REINFORCEMENT LEARNING

7.7.1 Introduction

The DAP framework connects the DP framework with the RL model. Both fields share many similarities, such as optimizing decision-making over time, but they also have differences in handling uncertainty and modeling environments. We can use the DAP framework to represent practical problems in both DP and RL, providing a unified approach to solving complex optimization and decision-making challenges. This connection is beneficial as we explore the automatic generation of algorithms.

RL is a framework designed to address practical problems represented as DAPs. It offers robust solutions, mainly when the environment's dynamics are unknown or too complex to model explicitly. While RL encompasses techniques conceptually related to DP, it extends beyond traditional DP by incorporating methods that derive the optimal policy through direct interaction with the environment. This interaction-based learning process can involve various algorithms, some rooted in DP principles. Consequently, RL provides a robust framework for tackling DAPs, which is especially useful in complex or unknown environments, a common scenario in real-world applications. RL enables learning without requiring a complete environment model by including DP-based algorithms and a broader range of techniques.

Imagine a robot navigating a grid to reach a specific goal while avoiding obstacles. In the DP approach, we assume the robot knows the exact layout of the grid, including the positions of all obstacles, the starting point, and the goal. The grid is divided into cells, and the robot can move up, down, left, or right. Each movement has a known probability of success and an associated cost or reward. The goal is to find the optimal path that minimizes the total cost or maximizes the total reward. DP uses a value function to evaluate the cost or reward of each cell in the grid and applies the Bellman equation to iteratively compute the optimal policy, which tells the robot the best move to make from each cell to reach the goal efficiently.

In contrast, the RL approach assumes the robot starts without knowledge of the grid layout, obstacle positions, or goal location. The robot learns to navigate the grid through trial and error by exploring different paths. Each time the robot takes an action (moving up, down, left, or right), it receives feedback in the form of rewards or penalties (e.g., a positive reward for moving closer to the goal or a negative reward for hitting an obstacle). Over time, the robot uses algorithms to update its understanding of the best actions to take from each cell based on the accumulated experiences. This way, the robot gradually learns an optimal grid navigation policy without needing a predefined environment model, making RL particularly useful in complex or unknown environments.

RL deals with problems where an agent interacts with an environment to learn an optimal policy for maximizing long-term rewards. Unlike traditional DP, which requires a complete model of the environment (transition and reward functions), RL algorithms operate in scenarios with partial or unknown models. These environments are often stochastic, meaning the outcome of an action is uncertain. Thus, we define the RL as

Given:

a) A set of states S that represents the agent's possible states. In RL, a state encapsulates all the information necessary to decide at a given time.

b) A set of actions A that are the agent's possible actions. The choice of action may depend on the current state.

c) A Transition Function T defines the probability of transitioning from one state to another, given a particular action. In RL, this function is denoted as $P(s' \mid s,a)$, which is the probability of transitioning to a state s' from state sss after taking action a.

d) A reward function R that gives the immediate reward received after transitioning from one state to another due to an action. It is denoted as $R(s,a,s')$.

e) Policy (π): A policy is a strategy that the agent follows, defined as a mapping from states to actions. Deterministic policies directly assign an action to each state; stochastic policies assign probabilities to each action for each state.

Then:

The objective is to maximize the expected return G from the initial state, s_0, where G is the sum of discounted rewards over time according to Equation 7.8, and the policy $\pi : S \to A$ maximizes the expected cumulative discounted reward over time.

This formal statement captures the essence of the RL problem in a mathematical framework, highlighting the key components of the state, action, transition dynamics, reward function, and the objective of maximizing the expected cumulative reward.

Example 7.6

Let's consider an autonomous vehicle needing to learn how to navigate efficiently in a city environment to reach a specified destination. The challenge is to minimize travel time while avoiding obstacles and adhering to traffic rules. First, we can establish a DAP formulation defining each necessary component.

- States: The state space includes the vehicle's location, speed, direction, and nearby obstacles or vehicles. Each state represents a unique configuration of these elements.
- Actions: Possible actions include accelerating, decelerating, turning (left or right), and maintaining current speed and direction.
- Transition Dynamics: The transition probabilities are initially unknown. They represent the likelihood of moving from one state to another, given an action influenced by factors like road conditions, traffic, and the vehicle's dynamics.
- Reward Function: The reward function assigns positive rewards for moving closer to the destination and negative rewards for undesirable outcomes like increased travel time, traffic rule violations, or collisions.
- Policy: The policy is a mapping from states to actions. The optimal policy maximizes the cumulative reward, which involves reaching the destination quickly and safely.

This DAP corresponds to an RL problem. The RL's suitability for this application stems from its adeptness in handling environments characterized by complexity, dynamism, and uncertainty. The vastness of the state and action spaces in urban driving scenarios makes the computational demands of conventional DP methods unfeasible. RL, however, thrives in such settings, owing to its capacity for optimization through direct interaction and its proficiency in making incremental improvements. This adaptability is advantageous for adjusting to the variable nature of road conditions and traffic patterns. As the RL agent engages with the environment over time, it progressively hones its policy, leading to increasingly effective decision-making. In this context, RL is a robust framework for addressing the autonomous vehicle navigation challenge, conceptualized as a DAP, and exemplifies the power of learning-based approaches in achieving optimal decision-making in complex, real-world scenarios.

7.7.2 RL Algorithms

This section aims to provide a brief and straightforward overview of key RL algorithms. These algorithms play a crucial role in solving deciding agent problems (DAPs) by enabling agents to learn optimal policies through interaction with their environments. While the descriptions here are concise, they highlight the fundamental principles and mechanisms that underpin each algorithm.

a) Q-learning. This fundamental RL algorithm enables an agent to learn the best actions to take in various states of an environment to maximize cumulative rewards. It operates model-free and doesn't require a predefined model of the environment's dynamics, such as transition probabilities and rewards. In Q-learning, the agent maintains a Q-values table, representing the expected utility of taking a given action in each state. The algorithm updates these Q-values as it explores the environment. It uses a simple formula that balances immediate rewards with estimated future rewards, adjusted by a learning rate and a discount factor. This update rule guides the agent to gradually converge on an optimal policy, which dictates the best action in each state.

   The learning process occurs over multiple episodes, where an episode represents a sequence of interactions from the initial state until a terminal state is reached (e.g., the agent reaches a goal or a maximum number of steps). During each episode, the agent makes decisions, observes rewards, and updates the Q-values based on experiences. The Q-Learning algorithm relies on a version of the Bellman equation to iteratively update the Q-values. Q-learning is particularly effective in environments with discrete states and actions and is well-suited for problems where the agent must learn from trial-and-error interactions with its environment.

   Applying Q-Learning to a DAP is essentially using it to learn the optimal policy for decision-making in a scenario where the agent must make a series of decisions to maximize some cumulative reward. The agent learns from its experiences by updating its Q-values, representing its current understanding of the best actions to take in each state to achieve its goal. Q-Learning is well-suited for problems where the environment's dynamics are unknown or complex, as it learns directly from environmental interactions. Thus, it is a powerful tool for solving DAPs, especially when traditional model-based approaches like DP are not feasible due to a lack of complete environmental knowledge.

b) Deep Q Networks (DQN). This algorithm represents an advanced form of Q-learning, integrating deep learning to handle complex, high-dimensional state spaces. In DQN, a neural network approximates the Q-value function, which traditionally would require a tabular approach. This network takes the current state of the environment as its input and outputs Q-values for each possible action, effectively estimating the potential future rewards for these actions. To enhance and stabilize the learning process, DQN incorporates techniques like experience replay, which allows the network to learn from past experiences stored in a replay buffer, and fixed Q-targets, which help mitigate the moving target problem inherent in iterative Q-value updates.

   The neural network in DQN is trained periodically rather than every time the agent takes an action to balance learning and computational efficiency. The network is trained using mini-batches of experience sampled from a replay buffer, which stores the agent's experiences (state, action, reward, next state). This training typically occurs after a fixed number of steps or episodes, where the agent uses its current policy to interact with the environment and collect experiences. Training the network involves adjusting weights to minimize the difference between the predicted and target Q-values, computed using the Bellman equation. A minimization nonlinear problem is solved to train the neural network to find the network's parameters. This optimization, in reality, estimates the Q-value function, providing the network with the ability to predict future rewards based on current states and actions. The training is computationally expensive, as it involves a minimization algorithm, but it doesn't need to happen at every step, which helps manage the computational load.

c) Policy Gradient Methods. These methods represent a distinct approach in RL, focusing directly on the policy function that maps states to actions. Unlike Q-learning, which centers on action values,

Policy Gradient methods optimize the policy. They achieve this by computing the gradient of expected rewards concerning the policy parameters and then adjusting these parameters in the direction that increases expected rewards. This method is particularly advantageous in scenarios with high-dimensional or continuous action spaces, where traditional value-based approaches might struggle. By learning a probability distribution over actions, Policy Gradient Methods can effectively handle complex decision-making tasks, offering more nuanced control over actions than methods that only learn deterministic policies. This approach is advantageous in environments where the action space is too large or continuous for value-based methods to be practical, allowing for more flexible and adaptive decision-making strategies.

In each episode of the policy gradient algorithm, the method focuses on solving an optimization problem where the variables are the parameters of the policy function. The primary objective is to optimize these parameters to maximize the cumulative reward the agent receives. Initially, the algorithm executes an episode involving the agent interacting with the environment and collecting data on states, actions, and rewards. This data collection phase effectively prepares the dataset for the subsequent optimization step. Once the episode is complete, the gradient algorithm uses the collected data to estimate the gradient of the expected return with respect to the policy parameters. This gradient then adjusts the policy parameters to increase the expected cumulative reward.

d) Actor-Critic Methods. These methods represent a hybrid approach that merges the concepts of policy gradient and value function optimization. These methods consist of two primary components: the Actor and the Critic. The Actor is responsible for updating the policy distribution and deciding which actions to take in given states. It does so based on feedback from the Critic, which estimates the value function, providing an assessment of the quality of the actions the Actor took. An advantage of this approach is its balance between direct policy optimization (Actor) and efficient value estimation (Critic). This balance helps mitigate the high variance often associated with standard Policy Gradient methods, leading to more stable and reliable learning. Actor-critic methods are particularly effective in environments where a balance of policy flexibility and value estimation is crucial, such as those with continuous action spaces or complex decision-making requirements.

The Actor-Critic methods solve two intertwined optimization problems. The Actor solves a policy optimization problem by adjusting the policy parameters to maximize the expected cumulative reward by resorting to the gradient ascent algorithm on the policy's parameters, where the gradient is estimated using the feedback from the Critic. On the other hand, the Critic solves a value function estimation problem, which involves minimizing the difference between the predicted value and the observed reward plus the discounted value of the next state. This last problem is a nonlinear minimization where the Critic's parameters are optimized to accurately reflect the value of the states under the current policy. The Actor and Critic are updated iteratively, where the Critic's accurate value estimates inform the Actor's policy updates, creating a synergistic optimization process. This dual optimization setup allows Actor-Critic methods to efficiently balance exploration and exploitation, making them robust for solving complex, high-dimensional RL problems.

A practical problem modeled as a DAP can be tackled effectively using RL methods. The fundamental optimization techniques are all four methods: Q-learning, Deep Q-Networks, Policy Gradient Methods, and Actor-Critic Methods. They involve solving complex optimization problems iteratively to find the best policies or value functions, ensuring that the agent makes optimal decisions in dynamic and uncertain environments.

7.7.3 Modeling the Automatic Generation of Algorithms as a DAP

To formulate the automatic generation of algorithms as a DAP within the context of the Master Problem presented in Chapter 2, we need to define the elements of a DAP: states, actions, rewards, and policies. The Master Problem involves finding the best solution for a given optimization problem by considering the spaces of algorithms, parameters, and instances. Let us consider the following definitions:

- States: In the context of the master problem, a state is a specific configuration of the algorithm's structure represented as a tree, parameter settings, and a set of problem instances. Each state encapsulates the current status of the algorithm generation process, including the functions and terminals selected so far and the parameters applied.

- Actions: They involve choosing the next step in the algorithm generation process. This selection could include choosing a new internal node to add to the tree, such as a loop, logical operator, or conditional. It may also involve selecting a terminal node, like a heuristic or exact method, or adjusting the parameters of the current nodes.
- Rewards: The reward function in this DAP considers the performance of the generated algorithm on the given problem instance. The reward is efficiency (e.g., computational time), effectiveness (e.g., solution quality), or a combination. The reward function incentivizes the generation of more effective and efficient algorithms.
- Policy: The policy in this DAP would be a strategy or a set of rules that guides the selection of actions (i.e., the development of the algorithm tree and parameter settings) based on the current state. The policy aims to maximize the expected reward, which translates to generating the most effective algorithm for the given problem instance.
- Objective: Maximize the overall performance of the generated algorithms across the given problem instances.

This formulation positions the automatic generation of algorithms as a fundamental problem within the DAP framework, where the agent (algorithm generator) iteratively improves its policy (algorithm structure and parameter settings) based on the feedback (performance) received from the environment (problem instances). The goal is to evolve toward an optimal policy that consistently generates high-performing algorithms for various problem instances.

Example 7.7

Define the essential components to construct algorithmic syntax trees that are useful for composing Binary Knapsack Problem (BKP) algorithms. In the BKP, several selected items are included in a knapsack with capacity $W > 0$. Let n be the number of items available, each one with a profit $p_j > 0$ and a weight $w_j > 0, j = 1,2,\ldots,n$. Additionally, consider that p_j, w_j, and W are integers. The mathematical formulation for this problem is discussed as a fundamental OR (Operational Research) problem in Chapter 3, Section 3.4. Thus, define internal nodes, terminal nodes, and the container for a BKP solution.

To construct algorithmic syntax trees useful for BKP algorithms, we must define the essential components: internal nodes, terminal nodes, and a container for a BKP solution. Let's consider the high-level programming functions found in most computer languages and defined by parameters P1 and P2, which may be functions or terminals. These are as follows:

- Loop While (P1, P2). P1 is the loop condition, and P2 is the loop expression. Thus, the loop controls the execution of P2, whereas P1 returns true; The expression returns The value of P2.
- And (P1, P2), which executes P1 and P2, returns true if both parameters are evaluated as True, returning false otherwise.
- Or (P1, P2) performs P2 only if P1 has a false value, and it returns true if P1 or P2 is true;
- Equal (P1, P2) returns true if P1 and P2 return the same value.
- If-then (P1, P2) performs P2 only if P1 returns a True value.
- If-then-else (P1, P2, P3) performs P2 if the performance of P1 returns a True value; otherwise, it performs P3 and returns the value P3 returns.

A container is an array specifying the items in the knapsack, i.e., the current solution. Furthermore, the set of terminals are functions specially designed for the BKP that select an element and insert or remove it from the knapsack container according to predefined criteria. Each terminal checks the feasibility of its operation, i.e., a terminal operates if and only if the move is feasible for the BKP. Specifically, five insertion and three deletion terminals are defined, each with a single criterion.

Insertion terminals. These terminals insert an item in the knapsack according to the maximum weight, the minimum weight, the maximum profit, or the maximum profit/weight ratio. Insertion terminals are specialized functions designed to add items to the knapsack (BKP container) according to specific criteria. These terminals ensure that each insertion is feasible, meaning that adding the item does not violate the knapsack's capacity constraint. The BKP container is an array representing the current set of selected items. Each insertion terminal evaluates all items and selects the one that best matches its criterion. Here are the five insertion terminals.

Insert Maximum Weight (Imax_weight):

- Function: Selects the item with the maximum weight that can still fit in the knapsack without exceeding capacity.
- Operation:
  - Iterate through all items not currently in the knapsack.
  - For each item, check if its weight, when added to the current total weight in the knapsack, does not exceed the knapsack capacity W.
  - Select the item with the highest weight to include feasibly.
  - Insert this item into the BKP container (the array storing the current solution).
  - Update the total weight and profit of the knapsack accordingly.

Insert Minimum Weight (Imin_weight):

- Function: Selects the item with the minimum weight that can still fit in the knapsack without exceeding capacity.
- Operation:
  - Iterate through all items not currently in the knapsack.
  - For each item, check if its weight, when added to the current total weight in the knapsack, does not exceed the knapsack capacity W.
  - Select the item with the lowest weight to include feasibly.
  - Insert this item into the BKP container.
  - Update the total weight and profit of the knapsack accordingly.

Insert Maximum Profit (Imax_profit):

- Function: Selects the item with the highest profit that can still fit in the knapsack without exceeding its capacity.
- Operation:
  - Iterate through all items not currently in the knapsack.
  - For each item, check if its weight, when added to the current total weight in the knapsack, does not exceed the knapsack capacity W.
  - Select the item with the highest profit among candidates to include feasibly.
  - Insert this item into the BKP container.
  - Update the total weight and profit of the knapsack accordingly.

Insert Minimum Profit (Imin_profit):

- Function: Selects the item with the lowest profit that can still fit in the knapsack without exceeding its capacity.
- Operation:
  - Iterate through all items not currently in the knapsack.
  - For each item, check if its weight, when added to the current total weight in the knapsack, does not exceed the knapsack capacity W.
  - Select the item with the lowest profit among candidates to include feasibly.
  - Insert this item into the BKP container.
  - Update the total weight and profit of the knapsack accordingly.

Insert Maximum Profit-to-Weight Ratio (Imax_profit_weight):

- Function: Selects the item with the highest profit-to-weight ratio that can still fit in the knapsack without exceeding capacity.
- Operation:
  - Iterate through all items not currently in the knapsack.
  - For each item, check if its weight, when added to the current total weight in the knapsack, does not exceed the knapsack capacity W.
  - Calculate the profit-to-weight ratio for each item.
  - Choose the item with the highest profit-to-weight ratio among those we can feasibly add.
  - Insert this item into the BKP container.
  - Update the total weight and profit of the knapsack accordingly.

Insert by Greedy Ratio (greedy_insert_by_ratio):

- Function: Selects items based on their profit-to-weight ratio and inserts them into the knapsack in descending order of their ratios, ensuring that they do not exceed the knapsack capacity.
- Operation:
  - Calculate Ratios, sort items in descending order, and iterate through all items.
  - Insert Items while the added weight is less than or equal to the capacity.
  - Insert each added item into the BKP container.

  - Update the current solution, the current total weight, and the profit of the knapsack accordingly.

Deletion terminals. These terminals remove an item according to the maximum weight, the minimum profit, or the minimum profit/weight. These terminals ensure each deletion is feasible, meaning removing the item does not violate the problem's constraints. Each deletion terminal evaluates all items currently in the knapsack and selects the one that best matches its criterion. Here are the three deletion terminals:

Delete Maximum Weight (Dmax_weight):

- Function: Selects the item with the maximum weight to remove from the knapsack.
- Operation:
  - Iterate through all items currently in the knapsack.
  - Identify the item with the highest weight.
  - Remove this item from the BKP container (the array storing the current solution).
  - Update the total weight and profit of the knapsack accordingly.

Delete Minimum Profit (Dmin_profit):

- Function: Selects the item with the minimum profit to remove from the knapsack.
- Operation:
  - Iterate through all items currently in the knapsack.
  - Identify the item with the lowest profit.
  - Remove this item from the BKP container.
  - Update the total weight and profit of the knapsack accordingly.

Delete Minimum Profit-to-Weight Ratio (Dmax_profit_weight):

- Function: Select the item with the lowest profit-to-weight ratio to remove from the knapsack.
- Operation:
  - Iterate through all items currently in the knapsack.
  - Calculate the profit-to-weight ratio for each item.
  - Identify the item with the lowest profit-to-weight ratio.
  - Remove this item from the BKP container.
  - Update the total weight and profit of the knapsack accordingly.

Example 7.8

Formulate the automatic generation of algorithms for the BKP as a DAP within the context of the Master Problem and consider the definitions in example 7.2. Define the DAP elements: states, actions, rewards, and policies.

To formulate the automatic generation of algorithms for the BKP as a Deciding Agent Problem within the context of the Master Problem and considering the definitions in Example 2, we need to define the DAP elements: states, actions, rewards, and policies.

- States: A state in this DAP is the current configuration of the algorithm's syntactic tree, comprising the functions and terminals, and the current solution to the BKP, which includes the items in the knapsack. Each state encapsulates:
  - The current structure of the algorithm.
  - The items currently selected in the knapsack.
  - The subsets of items available for future operations.
- Actions: Actions in this DAP involve choosing the next step in the algorithm generation process for the BKP. These actions include:
  - Choosing a new function to add to the tree, such as While, And, Or, Equal, If-Then, or If-Then-Else.
  - Choosing a terminal, such as insertion or deletion criteria.
  - Modifying the current tree structure by adding, removing, or changing nodes. Each action impacts the algorithm's structure and its approach to solving the BKP, influencing how the knapsack is filled and adjusted.
- Objective function: The function evaluates the performance of the generated algorithm on instances of the BKP and a penalty term. Such a function corresponds to the master problem objective function. Thus, instead of maximizing a reward, the objective is to minimize the average relative error between the algorithm's profit and the instance's optimal profit. This optimization occurs with the following measures:
  - The average relative error of the total profit of the items in the knapsack compared to the optimal profit for the BKP instances.

    - A penalty term that accounts for the relative error of the number of nodes in the syntactic tree versus the desired number of nodes. This penalty encourages the generation of concise algorithms.

    Thus, the objective function minimizes the average relative error in profit and the penalty for tree size and incentivizes the generation of algorithms that solve the BKP effectively and maintain computational efficiency and simplicity. By optimizing this objective function, the agent learns to produce high-quality and concise algorithms over multiple episodes.
- Policy: The policy is a strategy or set of rules that guides the selection of actions based on the current state. The policy aims to:
    - Minimize the cost function to generate the most effective algorithm for solving the BKP.
    - Adaptively choose actions that iteratively improve the algorithm's structure and performance. The policy is refined over time as the agent (algorithm generator) learns from the feedback provided by the environment (BKP instances).

In this formulation, the automatic generation of algorithms for the BKP corresponds to the DAP framework. The agent (algorithm generator) iteratively improves its policy based on the feedback from the environment (BKP instances), aiming to evolve toward an optimal policy that consistently generates high-performing algorithms for various BKP instances.

Solution strategy. To solve the automatic generation of algorithms for the BKP using RL, we consider an episode as the process of building a complete syntactic tree. This tree consists of internal nodes (functions like While, And, Or, Equal, If-Then, If-Then-Else) and leaf nodes (terminals that handle insertion and deletion criteria). The agent starts with an empty tree and selects actions to add nodes individually. An action involves adding a new function or a terminal to the tree. The agent continues to select actions until the tree is complete, meaning all leaf nodes are terminals, and internal nodes are connected functions. Each action updates the state, representing the tree's current structure and the knapsack's content.

The primary challenge in this agent formulation lies in the vast number of possible states. The current configuration of the algorithm syntactic tree and the current solution to the BKP characterize each state. The complexity arises from the combination of functions and terminals that can form the tree and the various items selected in the knapsack. This extensive state space can significantly complicate the learning process, making it computationally expensive and potentially infeasible to explore all states exhaustively.

The strategy to address this issue involves generating state categories based on the characteristics of the state. Instead of treating each unique state as distinct, we categorize similar states. These categories identify critical features of the syntactic tree (e.g., depth, number of nodes, number of types of nodes) and the solution container (e.g., total weight, total profit, number of items in the knapsack). The number of effective states is reduced by abstracting the states into broader categories, making the problem more tractable for RL algorithms. This approach leverages the generalization capability of the agent, enabling it to learn effective policies without needing to explore every possible state explicitly.

Example 7.9

Generate pseudo-code for a Q-learning algorithm to automatically create algorithms for the BKP using RL. The input consists of a BKP instances file containing values, weights, capacity, and optimal solutions for multiple instances, as well as parameters such as the number of episodes, learning rate (α), discount factor (γ), and exploration rate (ϵ). The output must be the best syntactic tree (algorithm) for solving the BKP and the learned Q-Table. The algorithm involves initializing the Q-learning environment, defining state categorization based on the number of nodes, branches, tree depth, profit blocks, and hits, and detailing the Q-learning steps to iteratively improve the policy for generating BKP algorithms.

The Algorithm 7.2, presents a pseudo-code for Q-learning to generate BKP algorithms. It involves initializing the Q-learning environment, generating an initial tree, categorizing states, selecting and applying actions, evaluating tree performance, updating the Q-table, and improving the policy iteratively for generating practical BKP algorithms. Algorithm 7.2 balances exploration and exploitation using an epsilon-greedy policy, adaptively updating the best syntactic tree based on the calculated rewards and minimizing the average error between the generated and optimal solutions. Algorithm 7.3 categorizes the states.

Given that the total number of states is unmanageable, it is necessary to establish state categories. Algorithm 7.3 categorizes the state of the current syntactic tree used for solving the BKP based on specific characteristics of the tree and the solution metrics. The input includes the current syntactic tree, characterized by the number of nodes (n), depth (d), and branches (b); the total instance profit accumulated (T_p); the optimal instance profit (T_p^*); and the number of near-optimal solutions found (*hits*). The output is an integer representing the categorized state (*Cat*). The algorithm begins by calculating the profit block (*pb*), which is the minimum between the ratio of the total profit to one-fourth of the optimal profit, and 3. Note that the block limits are $0 \le T_P < T_p^*/4$; $T_p^*/4 \le T_P < T_p^*/2$; $T_p^*/2 \le T_P < (4/3)T_p^*$ and $(4/3)T_p^* \le T_P$. It then calculates the hit category (h), which is the minimum between the number of hits and 4. The algorithm concludes by returning the categorized state (*Cat*).

The choice of weights c_0 *to* c_4 for the state categorization ensures that each feature contributes uniquely to the final state category. This approach avoids overlaps, maintains distinct state representations, and provides a structured way to categorize the state space, making it manageable for the Q-learning algorithm. One possibility for this set of numbers is to select multiples of 5 ($c_0 = 1, c_1 = 5, c_2 = 25, c_3 = 125, c_4 = 625$); thus, the formula ensures that changes in more significant features have a larger and distinct impact on the state category related to less significant features (like nodes and depth).

7.8 SUMMARY

This chapter delves into the intersection of DP and RL, emphasizing their combined potential in the automatic generation of algorithms. It begins with foundational concepts, tracing the evolution from Richard Bellman's early work to modern RL frameworks. The chapter explains how DP breaks complex decision-making processes into manageable subproblems and applies Bellman's principle of optimality to create efficient algorithms. RL extends these principles to dynamic and uncertain environments, leveraging machine learning advances. The main focus is framing real-world issues as DAP and suggesting a way to automatically develop algorithms for complex optimization tasks by viewing the overall problem through a dynamic lens. By combining DP and RL methods, the chapter shows how to create flexible, efficient algorithms for various optimization challenges, demonstrating the practical benefits of merging these two powerful computational approaches.

Algorithm 7.2 Q-learning to generate BKP algorithms.

Input:
```
BKP instances file contains values, weights,
 capacity, and
   the optimal solutions for several instances;
Set of functions and terminals;
Parameters: number of episodes, learning rate
 (α), discount factor (γ), and exploration rate
 (ε), decay_rate (d_r);
```

Output:
```
Best syntactic tree (algorithm) for solving the
 BKP;
Q-Table representing the learned policy;
Reward progression over episodes;
```

Begin
```
Initialize Q-table Q(S, A) to zeros;
For each episode in number_of_episodes:

  ε = max( ε_min, ε * d_r);

 Generate an initial tree with random depth and
 structure;
 While the tree structure is not valid (too many
 nodes or too deep):
   Generate a new tree;
 Initialize state s using the categorize_state
 algorithm with initial values;
 Repeat until the tree is complete:
   Choose action a using epsilon-greedy policy
     based on state s
    Evaluate tree performance with old_avg_error;
   Apply action a on the tree:
    If a = 0: Replace a function in the tree;
    If a = 1: Replace a terminal in the tree;
    If a = 2: Shrink the tree;
    If a = 3: Add a random subtree to the tree;
    If a = 4: Swap subtrees with another randomly
      generated tree;
    If a = 5: Mutate a subtree;
```

```
    Evaluate new tree performance with new_avg_
      error;
    Calculate reward r as old_avg_error - new_avg_
      error;
    If new_avg_error < best_error Then Update best_
      error and best_tree;
    Update accumulated_profit and optimal_profit;
    Update hits based on solutions meeting near-
      optimal criteria;
    Categorize next_state using updated values;
    Update Q-table Q(S, A) using Bellman equation:
    Set state S to next_state;
End.
```

Algorithm 7.3 Categorize_State.

Input:
```
 n: Number of nodes in the current syntactic tree
 d: Depth of the current syntactic tree
 b: Number of branches in the current syntactic tree
 Tp: Total profit accumulated for the instance
 Tp*: Optimal profit of the instance
 hits: Number of near-optimal solutions found
```

Output:
```
 Cat: Integer representing the categorized state;
```

Begin

pb = min $(T_P/(T_p^*/4), 3)$ #profit block

h = min $(hits, 4)$ #hit category

$Cat = c_0 n + c_1 d + c_2 b + c_3 pb + c_4 h$ #Categorized state

Return Cat;

End

EXERCISES

1. Explain Bellman's principle of optimality and its significance in DP. Provide a detailed example of how to apply this principle to a simple shortest-path problem in a directed graph.

2. Apply RL directly to solve a specific instance of the (BKP). The goal is not to generate an algorithm for solving BKP but to use RL to find the optimal solution for a given BKP instance. Define this challenge as a DAP identifying states, actions, and rewards for the RL agent.

3. Formulate the binary knapsack problem and solve it using DP. Show step-by-step calculations and explain the process of building the solution table.

4. Using the state categorization function discussed in the chapter, categorize a given syntactic tree for the binary knapsack problem. Assume specific values for the number of nodes, depth, branches, accumulated profit, optimal profit, and hits. Explain each step of the categorization process.

5. Implement the pseudo-code for the Q-learning algorithm provided in the chapter to generate algorithms for the multidimensional knapsack problem (MKP). The MKP is a generalization of the binary knapsack problem with multiple constraints (dimensions). Test your implementation with a small set of MKP instances and report the results, including the best syntactic tree and rewards progression over episodes.

6. Extend the concept of modeling the automatic generation of algorithms as a deciding agent problem to the Traveling Salesman Problem (TSP). The TSP involves finding the shortest possible route that visits each city exactly once and returns to the origin city.

7. Following the procedure in this chapter, use the criteria to define terminals oriented to generate metaheuristics as defined in Section 6.5. Extend the approach to automatically generate an algorithm for solving the master problem to generate hybrid metaheuristics. Particularize the procedure to generate a metaheuristic for the graph coloring problem, another challenging combinatorial optimization problem.

8. Focus on tuning the parameters of the RL algorithm described in Example 7.4. Identify all the relevant parameters in the RL algorithm and develop a pseudo-code for tuning these parameters using a set of BKP instances.

CHAPTER 8

Conclusions and Future Trends

8.1 INTRODUCTION

As we conclude this exploration into the AGA, reflecting on the journey unfolding throughout this book is crucial. Starting with the foundational principles outlined in Chapters 1 and 2, we have delved into the intricate relationship between optimization, combinatorial problems, and the process of algorithmic creation. These early chapters laid the groundwork for understanding how AGA operates as a technique to automate the discovery of practical algorithms, particularly in solving NP-hard combinatorial optimization problems. We proposed new avenues for creating efficient and tailored algorithms for specific problem instances by framing algorithm generation as an optimization meta-problem.

Chapters 3–7 further expanded on these foundations, exploring AGA's practical applications and theoretical underpinnings in greater detail. In Chapter 3, we addressed the challenges of modeling real-world problems for AGA, emphasizing the importance of precise problem representation to ensure the generated algorithms are effective. Chapter 4 introduced genetic programming and other evolutionary algorithms as powerful tools in the AGA process, demonstrating how these methods can explore vast solution spaces to optimize algorithm design. The subsequent chapters, particularly Chapters 5 and 6, highlighted the integration of machine learning and metaheuristics with AGA, showcasing the potential for these techniques to enhance algorithm performance and adaptability.

DOI: 10.1201/9781003542896-8

The discussion of reinforcement learning in Chapter 7 provided additional insights into constructing algorithms through interaction with the problem instances over time. This concept, when combined with AGA, presents a framework for developing algorithms that can evolve and adapt to changing problem landscapes, further pushing the boundaries of what is possible in computational problem-solving.

AGA stands at the forefront of algorithmic innovation. The rapid advancements in AI and optimization technologies present opportunities and challenges for the future of AGA. While the techniques and methodologies discussed throughout this book provide a solid foundation, it is clear that the field is dynamic and will continue to evolve. In this closing discussion, we synthesize the knowledge gathered throughout the book, drawing connections between the various concepts, techniques, and applications explored. We also identify emerging trends, potential innovations, and the challenges ahead. By doing so, we aim to provide a comprehensive view of the current state of AGA while offering a vision for its future evolution, ensuring that it remains a dynamic field in the years to come.

8.2 FUTURE DIRECTIONS AND RESEARCH PATHS

Driven by optimization techniques, artificial intelligence, and computational power advancements, AGA can evolve rapidly. As we look to the future, several promising research paths and potential developments emerge that could enhance the capabilities and applications of AGA.

One of the most exciting future directions lies in the continued integration of AGA with cutting-edge AI technologies. As AI advances, particularly in machine learning and deep learning, there is a growing opportunity to harness these techniques to improve algorithm generation. Future research could explore how AGA can leverage AI-driven models to predict the performance of candidate algorithms more accurately, reducing the computational effort required for optimization. Additionally, as AI models become more sophisticated, there is potential for developing algorithms capable of solving even more complex and dynamic problems than those currently addressed by AGA.

While AGA has primarily focused on combinatorial optimization problems, there is a vast potential for expanding its scope into new and emerging domains. Future research could investigate the application of AGA in fields such as personalized medicine, where algorithms could be automatically generated to tailor treatments to individual patients based on their unique genetic profiles. Similarly, in autonomous systems, AGA could be employed to develop algorithms that optimize decision-making processes in real-time, enhancing the performance and safety of autonomous vehicles and robots. Exploring these new applications could broaden the impact of AGA and lead to the discovery of novel algorithmic strategies that push the boundaries of current technology.

Another promising avenue for future research is the development of hybrid models that combine AGA with other optimization and AI paradigms. For instance, integrating parallelism with AGA could open new possibilities for solving previously intractable problems. Additionally, cross-disciplinary approaches that integrate insights from biology, economics, and social sciences with AGA have the potential to yield innovative solutions to complex real-world challenges. AGA's ability to personalize algorithms for individual devices also presents opportunities in cybersecurity, enabling the development of tailored algorithms that can detect and prevent fraud and protect against information theft. By exploring these hybrid and interdisciplinary approaches, researchers can develop more robust, versatile, and secure algorithms capable of addressing a broader range of problems.

As AGA becomes increasingly sophisticated, it is essential to consider the ethical implications of automatically generated algorithms. Ensuring that these algorithms are transparent, interpretable, and aligned with ethical standards will be a critical area of focus for future research. There is a growing need for methods that generate efficient algorithms and provide insights into their decision-making processes, allowing users to understand and trust the outcomes. Future work could explore techniques for embedding ethical considerations into the AGA process, ensuring that generated algorithms are fair, unbiased, and aligned with societal values.

The implications of NP-completeness are profound for AGA. Since many combinatorial optimization problems tackled by AGA fall into this category, the automatic generation of algorithms must contend with the inherent difficulty of these problems. Researchers have developed various heuristics and approximation methods to address NP-complete problems, but the computational complexity persists as a barrier, especially as problem instances grow more extensive and intricate. However, AGA offers a unique advantage in the search for efficient solutions by allowing us to explore vast algorithmic spaces. This capability opens up the possibility of discovering polynomial-time algorithms for NP-complete problems.

As we consider AGA's future directions and research paths, integrating emerging technologies like LLMs is pivotal. The emergence of LLMs presents a game-changing opportunity for AGA. With the advanced capabilities of LLMs, generating the computer code necessary to implement AGA, whether through genetic programming or other metaheuristic algorithms, has become significantly more accessible. This development simplifies the construction of automated algorithm generators tailored to various problems. Additionally, LLMs facilitate the combination of well-known algorithms, enabling the creation of new algorithms that integrate the most effective components of each. This synergy between LLMs and AGA enhances the potential for innovation in algorithm design, making it easier to address complex challenges across diverse domains.

8.3 CLOSING THOUGHTS

The synergy between AGA and the emerging LLMs represents a significant development in computational problem-solving, with potential implications for how algorithms are generated, optimized, and applied across various domains. LLMs can parse large volumes of research, extract insights, and suggest algorithmic components that AGA can further explore and optimize, thereby accelerating the generation process and aligning it with the latest advancements in the field. LLMs also play a critical role in designing objective functions and constraints within AGA, providing suggestions based on historical data and domain-specific knowledge.

The combination of AGA and LLMs enables adaptive learning systems to evolve. LLMs can provide real-time feedback on algorithm performance, guiding refinements and ensuring algorithms remain effective as problem contexts change. By learning from past experiences and extracting patterns from previous generations, LLMs help AGA develop finely tuned algorithms for specific domains.

Current advances in AI, coupled with the insights provided in this book, equip readers with the tools and knowledge needed to construct algorithm generator machines tailored to specific problems rapidly. By leveraging the power of metaheuristics and the modular approach to algorithm design outlined here, one can combine various optimization techniques or elementary components commonly found in effective algorithms to create customized solutions. This approach allows for the automated generation of efficient and adaptable algorithms to diverse problem contexts. With the practical guidance and examples provided in this book, readers are encouraged to apply these methods, experiment with different combinations, and begin constructing their algorithm generators to tackle real-world challenges. Now is the time to put theory into practice and explore AI and AGA's vast possibilities for solving complex problems.

Bibliography

Acevedo, N., Rey, C., Contreras-Bolton, C., & Parada, V. (2020). Automatic Design of Specialized Algorithms for the Binary Knapsack Problem. *Expert Systems with Applications*, 141, 112908.

Agapitos, A., & O'Neill, M. (2021). *Genetic Programming Techniques: Building Artificial Minds*. Chapman and Hall/CRC.

Ahuja, R. K., Magnanti, T. L., & Orlin, J. B. (1993). *Network Flows: Theory, Algorithms, and Applications*. Prentice Hall.

Alba, E., & Martí, R. (2004). *Metaheuristic Procedures for Training Neural Networks*. Operations Research/Computer Science Interfaces Series, 36. Springer New York, NY.

Alpaydin, E. (2020). *Introduction to Machine Learning* (4th ed.). The MIT Press.

Applegate, D., Bixby, R., Chvátal, V., & Cook, W. (2006). *The Traveling Salesman Problem: A Computational Study*. Princeton University Press.

Banzhaf, W., Nordin, P., Keller, R. E., & Francone, F. D. (1998). *Genetic Programming: An Introduction*. Morgan Kaufmann.

Barros, R. C., De Carvalho, A. C. P. L. F., & Freitas, A. A. (2015). *Automatic Design of Decision-Tree Induction Algorithms*. Springer International Publishing.

Bazaraa, M. S., Jarvis, J. J. & Sherali, H. D. (2009). *Linear Programming and Network Flows* (4o ed.). Wiley.

Bazaraa, M. S., Sherali, H. D., & Shetty, C. M. (2013). *Nonlinear Programming: Theory and Algorithms* (3rd ed.). Wiley.

Bellman, R. (1957). *Dynamic Programming*. Princeton University Press.

Bellman, R. (1962). Dynamic Programming Treatment of the Travelling Salesman Problem. *Journal of the ACM*. https://doi.org/10.1145/321105.321111

Bello, I., Pham, H., Le, Q. V., Norouzi, M., & Bengio, S. (2016). Neural Combinatorial Optimization With Reinforcement Learning. arXiv preprint arXiv:1611.09940

Bengio, Y., Lecun, Y., & Hinton, G. (2015). Deep Learning. *Nature*, 521, 436–444.

Bertolini, V., Rey, C., Sepulveda, M. & Parada, V. (2018). Novel Methods Generated by Genetic Programming for the Guillotine-Cutting Problem. *Scientific Programming*, 2018, e6971827.

Bertsekas, D. P. (2016). *Nonlinear programming* (3rd ed.). Athena Scientific.

Bertsekas, D. P., & Tsitsiklis, J. N. (1996). *Neuro-Dynamic Programming*. Athena Scientific.

Bischl, B., Binder, M., Lang, M., Pielok, T., Richter, J., Coors, S., Thomas, J., Ullmann, T., Becker, M., Boulesteix, A.-L., Deng, D., & Lindauer, M. (2021).

Hyperparameter Optimization: Foundations, Algorithms, Best Practices, and Open Challenges. arXiv preprint arXiv:2107.058476.

Bishop, C. M. (2006). *Pattern Recognition and Machine Learning*. Springer.

Blum, C., & Roli, A. (2003). Metaheuristics in Combinatorial Optimization: Overview and Conceptual Comparison. *ACM Computing Surveys*, 35(3), 268–308.

Boyd, S., & Vandenberghe, L. (2004). *Convex Optimization*. Cambridge University Press.

Branke, J., Corne, D., Deb, K. (Eds.). (2023). *Multi-objective Genetic Programming: New Approaches and Applications*. Springer. This book delves into the use of genetic programming for solving multi-objective optimization problems.

Breiman, L. (2001). Random Forests. *Machine Learning*, 45(1), 5–32.

Brockman, G., et al. (2016). OpenAI Gym. arXiv preprint arXiv:1606.01540

Brownlee, J. (2020). *Generative Adversarial Networks with Python: Deep Learning Generative Models for Image Synthesis and Image Translation*. Machine Learning Mastery.

Burke, E., Hyde, M., Kendall, G., & Woodward, E. J. (2007). *Automatic Heuristic Generation With Genetic Programming: Evolving a Jack-of-All-Trades or a Master of One*. In Proceedings of the 2007 IEEE Congress on Evolutionary Computation. IEEE. pp. 1559–1565.

Burke, E. K., Hyde, M. R., Kendall, G., & Woodward, J. (2012). Automating the Packing Heuristic Design Process With Genetic Programming. *Evolutionary Computation*, 20(1):63–89.

Buyya, R., Broberg, J., & Goscinski, A. (2016). *Big Data: Principles and Paradigms*. Morgan Kaufmann.

Chollet, F. (2021). *Deep Learning With Python* (2nd ed.). Manning Publications.

Chvatal, V. (1983). *Linear Programming* (1st ed.). W.H.Freeman & Co Ltd.

Contreras-Bolton, C., Gatica, G. & Parada, V. (2013). Automatically Generated Algorithms for the Vertex Coloring Problem. *PLoS ONE*, 8(3), e58551.

Cook, W. (2012). *In Pursuit of the Traveling Salesman: Mathematics at the Limits of Computation*. Princeton University Press.

De Jong, K. A. (2006). *Evolutionary Computation: A Unified Approach*. MIT Press.

Deb, K. (2001). *Multi-objective Optimization Using Evolutionary Algorithms*. Wiley.

Deisenroth, M. P., Neumann, G., & Peters, J. (2013). A Survey on Policy Search for Robotics. *Foundations and Trends in Robotics*, 2(1–2), 1–142.

Dorigo, M., & Stützle, T. (2004). *Ant Colony Optimization*. MIT Press.

Drake, J. H., Kheiri, A., Özcan, E. & Burke, E. K. (2020). Recent Advances in Selection Hyper-Heuristics. *European Journal of Operational Research*, 285(2), 405–428.

Eiben, A. E., Michalewicz, Z., Schoenauer, M., & Smith, J. E. (2007). Parameter Control in Evolutionary Algorithms. In F. G. Lobo, C. F. Lima, E. Z. Michalewicz (Eds.), *Parameter Setting in Evolutionary Algorithms, Studies in Computational Intelligence*, pp. 19–46. Springer.

Eiben, A. E., & Smith, J. E. (2015). *Introduction to Evolutionary Computing* (2nd ed.). Springer.

Eiselt, H. A., & Sandblom C. L. (2010). *Integer Programming and Network Models. Softcover Reprint of Hardcover* (1st ed). Springer.

Etter, D. M., Ingber, J. E., & Recktenwald, G. W. (2016). *Engineering Problem Solving With C++* (4th ed.). Pearson.

Fogel, D. B. (2006). *Evolutionary Computation: Toward a New Philosophy of Machine Intelligence*. IEEE Press.

Forsyth, R. (1981). *BEAGLE A Darwinian Approach to Pattern Recognition*. Kybernetes.

François-Lavet, V., Henderson, P., Islam, R., Bellemare, M. G., & Pineau, J. (2018). An Introduction to Deep Reinforcement Learning. *Foundations and Trends in Machine Learning*, 11. https://doi.org/10.1561/2200000071

Gandomi, A., Alavi, A. H., Ryan, C., & Shirkhorshidi, A. S. (Eds.). (2019). *Data Analytics in Engineering*. Elsevier.

Garey, M. R., & Johnson, D. S. (1979). *Computers and Intractability: A Guide to the Theory of NP-Completeness*. W. H. Freeman.

Garfinkel, R. S., & Nemhauser, G. L. (2012). *Integer Programming*. Dover Publications.

Gass, S. (2010). *Linear Programming: Methods and Applications* (5th ed.). Dover Publications.

Geem, Z.W., Kim, J.H., & Loganathan, G.V. (2001). A New Heuristic Optimization Algorithm: Harmony Search. *Simulation*, 76(2), 60–68.

Gendreau, M., & Potvin, J. (2010). *Handbook of Metaheuristics* (2nd ed.). Springer.

Géron, A. (2019). *Hands-On Machine Learning With Scikit-Learn, Keras, and TensorFlow: Concepts, Tools, and Techniques to Build Intelligent Systems* (2nd ed.). O'Reilly Media.

Glover, F. (1986). Future Paths for Integer Programming and Links to Artificial Intelligence. *Computers & Operations Research*, 13(5), 533–549.

Glover, F., & Kochenberger, G. A. (Eds.). (2003). *Handbook of Metaheuristics*. Springer.

Glover, F., & Laguna, M. (1997). *Tabu Search*. Kluwer Academic Publishers.

Goldberg, D. E. (1989). *Genetic Algorithms in Search, Optimization, and Machine Learning*. Addison-Wesley.

Gonzalez, T. F. (2007). *Handbook of Approximation Algorithms and Metaheuristics* (1o ed). Chapman and Hall/CRC.

Goodfellow, I., Bengio, Y., & Courville, A. (2016). *Deep Learning*. MIT Press.

Harik, G. R., Cantú-Paz, E., Goldberg, D. E., & Miller, B. L. (1999). The Gambler's Ruin Problem, Genetic Algorithms, and the Sizing of Populations. *Evolutionary Computation*, 7, 231–253.

Hastie, T., Tibshirani, R., & Friedman, J. (2009). *The Elements of Statistical Learning: Data Mining, Inference, and Prediction*. Springer.

Hillier, F. S., & Lieberman, G. J. (2015). *Introduction to Operations Research* (10th ed.). McGraw-Hill Education.

Holland, J. H. (1975). *Adaptation in Natural and Artificial Systems*. University of Michigan Press.

Hoos, H. H., & Stützle, T. (2004). *Stochastic Local Search: Foundations & Applications*. Morgan Kaufmann.

Hughes, M., Goerigk, M. & Dokka, T. (2021). Automatic Generation of Algorithms for Robust Optimisation Problems Using Grammar-Guided Genetic Programming. *Computers & Operations Research*, 133, 105364.

Hutter, F., Hoos, H. H., Leyton-Brown, K., & Stützle, T. (2009). ParamILS: An Automatic Algorithm Configuration Framework. *Journal of Artificial Intelligence Research*, 36(1), 267–306.

Iba, H., & Noman, N. (Eds.). (2022). *Field Guide to Genetic Programming With Python*. Springer. A practical guide to applying genetic programming in Python, including numerous examples and case studies.

Iturra, S., Contreras-Bolton, C. & Parada, V. (2022). Automatic Generation of Metaheuristic Algorithms. In B. Dorronsoro, F. Yalaoui, E.-G. Talbi, & G. Danoy (Eds.), *Metaheuristics and Nature Inspired Computing*, pp. 48–58. Springer International Publishing.

Kaelbling, L. P., Littman, M. L., & Moore, A. W. (1996). Reinforcement Learning: A Survey. *Journal of Artificial Intelligence Research*, 4, 237–285.

Karp, R. M. (1972). Reducibility Among Combinatorial Problems. In R. E. Miller & J. W. Thatcher (Eds.), *Complexity of Computer Computations* (pp. 85–103). Springer.

Kennedy, J., & Eberhart, R.C. (1995). Particle Swarm Optimization. In Proceedings of the IEEE International Conference on Neural Networks, pp. 1942–1948.

Kirkpatrick, S., Gelatt, C.D., & Vecchi, M.P. (1983). Optimization by Simulated Annealing. *Science*, 220(4598), 671–680.

Knuth, D. E. (1998). *The Art of Computer Programming. Sorting and Searching*, Vol. 3, 2nd Edition. Addison-Wesley.

Kocsis, L., & Szepesvári, C. (2006). Bandit based Monte-Carlo Planning. In European Conference on Machine Learning.

Kocsis, Z. A., & Swan, J. (2018). Genetic Programming Proof Search Automatic Improvement. *Journal of Automated Reasoning*, 60(2), 157–176.

Konda, V. R., & Tsitsiklis, J. N. (2000). Actor-Critic Algorithms. In *Advances in Neural Information Processing Systems*, 13, 1008–1014.

Kossiakoff, A., Sweet, W. N., Seymour, S. J., & Biemer, S. M. (2011). *Systems Engineering Principles and Practice* (2nd ed.). Wiley.

Koza, J. (2003). *Genetic Programming IV: Routine Human-Competitive Machine Intelligence*. Kluwer Academic Pub.

Koza, J., &Poli, R. (2005). *Genetic Programming*. Massachusetts Institute of Technology.

Koza, J. R. (1992). *Genetic Programming: On the Programming of Computers by Means of Natural Selection*. MIT Press.

Koza, J. R. (1994). *Genetic Programming II: Automatic Discovery of Reusable Programs*. The MIT Press.

Krasnogor, N., & Smith, J. (2005). A Tutorial for Competent Memetic Algorithms: Model, Taxonomy, and Design Issues. *IEEE Transactions on Evolutionary Computation*, 9(5), 474–488.

Lawler, E. L., Lenstra, J. K., Rinnooy Kan, A. H. G., & Shmoys, D. B. (Eds.). (1985). *The Traveling Salesman Problem: A Guided Tour of Combinatorial Optimization*. Wiley.

Lillicrap, T. P., et al. (2015). Continuous Control With Deep Reinforcement Learning. arXiv preprint arXiv:1509.02971.

López-Ibáñez, M., Dubois-Lacoste, J., Pérez Cáceres, L., Birattari, M., & Stützle, T. (2016). The Irace package: Iterated Racing for Automatic Algorithm Configuration. *Operations Research Perspectives*, 3, 43–58.

Loyola, C., Sepúlveda, M., Solar, M., Lopez, P., & Parada, V. (2016). Automatic Design of Algorithms for the Traveling Salesman Problem. *Cogent Engineering*, 3(1), 1255165.

Luke, S. (2000). Genetic Programming: Producing Competitive Algorithms. PhD thesis. University of Maryland, College Park.

McDermott, J., Castelli, M., Sekanina, L., Haasdijk, E., & García-Sánchez, P. (Eds.). (2020). Genetic Programming: In 23rd European Conference, EuroGP 2020. Springer. It presents research papers from one of the premier conferences on genetic programming.

Michalewicz, Z., & Fogel, D. B. (2004). *How to Solve It: Modern Heuristics* (2nd ed.). Springer.

Mitchell, M. (1998). *An Introduction to Genetic Algorithms*. MIT Press.

Mnih, V., et al. (2013). Playing Atari With Deep Reinforcement Learning. arXiv preprint arXiv:1312.5602.

Mnih, V., et al. (2015). Human-Level Control Through Deep Reinforcement Learning. *Nature*, 518(7540), 529–533.

Murphy, K. P. (2012). *Machine Learning: A Probabilistic Perspective*. MIT Press.

Nakayama, A. (2014). *Practical Engineering Problem Solving Techniques*. Butterworth-Heinemann.

Nocedal, J., & Wright, S. J. (2006). *Numerical Optimization* (2nd ed.). Springer.

Oberg, J., Das, S., Shekhar, R., & Feng, M. (2013). *Model-Based Engineering for Complex Electronic Systems*. Elsevier.

Papadimitriou, C. H., & Steiglitz, K. (1998). *Combinatorial Optimization: Algorithms and Complexity*. Dover Publications.

Parada, L., Herrera, C., Sepúlveda, M., & Parada, V. (2016). Evolution of New Algorithms for the Binary Knapsack Problem. *Natural Computing*, 15(1), 181–193.

Piotrowski, A.P. (2017). Review of Differential Evolution Population Size. *Swarm and Evolutionary Computation*, 32, 1–24.

Poli, R., Langdon, W. B., & McPhee, N. F. (2008). A Field Guide to Genetic Programming. *Lulu.com*.

Powell, W. B. (2011). *Approximate Dynamic Programming: Solving the Curses of Dimensionality* (2nd ed.). Wiley.

Provost, F., & Fawcett, T. (2013). *Data Science for Business*. O'Reilly Media.

Rong, G., Huang, T., & Xu, Z. (2020). A Review on Automatic Machine Learning Technology and Its Prospects. *Proceedings of the IEEE*, 108(8), 1–20.

Russell, S., & Norvig, P. (2020). *Artificial Intelligence: A Modern Approach* (4th ed.). Pearson.

Ryser-Welch, Patricia, Miller, Julian F., & Asta, Shahriar (2015). Generating Human-Readable Algorithms for the Travelling Salesman Problem Using Hyper-Heuristics. In Proceedings of the Companion Publication of the 2015

Annual Conference on Genetic and Evolutionary Computation, GECCO Companion '15, pp. 1067–1074. Association for Computing Machinery.

Schmidhuber, J. (2015). Deep Learning in Neural Networks: An Overview. *Neural Networks*, 61, 85–117.

Silver, D., Huang, A., Maddison, C. J., Guez, A., Sifre, L., Van Den Driessche, G., … & Hassabis, D. (2016). Mastering the Game of Go With Deep Neural Networks and Tree Search. *Nature*, 529(7587), 484–489.

Singh, S., & Reddy, C. K. (Eds.). (2016). *Big Data Analytics: Methods and Applications*. Springer.

Sipper, M., Fu, W., Moore, J. H. (Eds.). (2018). *Handbook of Genetic Programming Applications*. Springer.

Skinner, B. F. (1953). *Science and Human Behavior*. Macmillan.

Sörensen, K., & Glover, F. (2013). Metaheuristics. *Encyclopedia of Operations Research and Management Science*, pp. 960–970. https://doi.org/10.1007/978-1-4419-1153-7

Spector, L., Langdon, W. B., O'Reilly, U.-M., & Angeline, P. J. (Eds.). (1999). *Advances in Genetic Programming 3*. MIT Press.

Sutton, R. S., Barto, A. G. (2020). *Reinforcement Learning: An Introduction* (2nd ed.). The MIT Press.

Taha, H. A. (2017). *Operations Research: An Introduction* (10th ed.). Pearson.

Talbi, E. G. (2009). *Metaheuristics: From Design to Implementation*. Wiley.

Turing, A. M. (1950). Computing Machinery and Intelligence. *Mind*, LIX(236), 433–460.

van Leeuwen, J. (Ed.). (1998). *Handbook of Theoretical Computer Science, Vol. A, Algorithms and Complexity*. MIT Press.

Vose, M. D. (1999). *The Simple Genetic Algorithm: Foundations and Theory*. MIT Press.

Watkins, C. J. C. H. (1989). Learning from Delayed Rewards. PhD Thesis. University of Cambridge.

Watkins, C. J. C. H., & Dayan, P. (1992). Q-learning. *Machine Learning*, 8(3–4), 279–292.

Whigham, P. A., Dick, G., Maclaurin, J. (2019). *Genetic Programming Theory and Practice XVI*. Springer.

Williams, H. P. (1999). *Model Building in Mathematical Programming* (4th ed.). Wiley.

Winston, W. L. (2004). *Operations Research: Applications and Algorithms* (4th ed.). Brooks/Cole.

Wolpert, D.H., & Macready, W.G. (1997). No Free Lunch Theorems for Optimization. *IEEE Transactions on Evolutionary Computation*, 1(1), 67–82.

Wolsey, L. A., & Nemhauser, G. L. (1999). *Integer and Combinatorial Optimization* (1o ed.). Wiley-Interscience.

Yang, Xin-She. (2008). *Nature-Inspired Metaheuristic Algorithms*. Luniver Press.

Yang, Xin-She (2010). *Nature-Inspired Metaheuristic Algorithms*. Luniver Press.

Yang, Xin-She, & Deb, S. (2009). Cuckoo Search via Lévy Flights. In World Congress on Nature & Biologically Inspired Computing (NaBIC), pp. 210–214.

Index

For Product Safety Concerns and Information please contact our EU representative GPSR@taylorandfrancis.com
Taylor & Francis Verlag GmbH, Kaufingerstraße 24, 80331 München, Germany

www.ingramcontent.com/pod-product-compliance
Ingram Content Group UK Ltd.
Pitfield, Milton Keynes, MK11 3LW, UK
UKHW011010180926
475701UK00004B/663
* 9 7 8 1 0 3 2 8 9 1 3 8 5 *